融合之间　　转型中的　　当代中国建筑

Fusion and Harmony
Contemporary Chinese Architecture in the Process of Transition

中国建筑学会 ｜ 中国建筑学会建筑师分会 ｜ 华东建筑集团股份有限公司　编著

中国建筑工业出版社

图书在版编目（CIP）数据

融合之间：转型中的当代中国建筑 / 中国建筑学会，中国建筑学会建筑师分会，华东建筑集团股份有限公司编著.
—北京：中国建筑工业出版社，2017.8
ISBN 978-7-112-21122-7

Ⅰ.①融… Ⅱ.①中… ②中… ③华… Ⅲ.①建筑设计—作品集—中国—现代 Ⅳ.① TU206

中国版本图书馆CIP数据核字（2017）第 203782 号

责任编辑：陆新之　张　明　段　宁
责任校对：焦　乐　张　颖

融合之间——转型中的当代中国建筑
中国建筑学会
中国建筑学会建筑师分会　　编著
华东建筑集团股份有限公司

*

中国建筑工业出版社出版、发行（北京海淀三里河路9号）
各地新华书店、建筑书店经销
北京雅昌艺术印刷有限公司制版印制

*

开本：787×1092 毫米　1/12　印张：27　字数：400千字
2017年8月第一版　2017年8月第一次印刷
定价：198.00元
ISBN 978-7-112-21122-7
　　　　（30774）

版权所有　翻印必究
如有印装质量问题，可寄本社退换
（邮政编码 100037）

Sponsor: Architectural Society of China

Organizer: Architects Branch of Architectural Society of China
　　　　　 Arcplus Group PLC
　　　　　 Beijing Archicity Consulting Co., Ltd

Preparatory Committee: Xiu Long, Zhang Baiping, Gu Yongxin, Cui Kai, Meng Jianmin, Zhuang Weimin, Shao Weiping, Shen Di, Li Cundong, Qian Fang, Xue Ming, Sun Zonglie, Fu Shaohui, Gao Wenyan

Chief Curator: Shen Di

The Executive Curator: Gao Wenyan, Huang Qiuping, Jiang Shifeng, DingShun, Dong Yi, Zhou Xiaowen

Supporters: Arcplus Group PLC
　　　　　 Beijing Institute of Architectural Design Co.,Ltd
　　　　　 China Architecture Design Group
　　　　　 China IPPR International Engineering Co., Ltd.

主办单位：中国建筑学会
承办单位：中国建筑学会建筑师分会
　　　　　华东建筑集团股份有限公司
　　　　　北京凯欣城市发展咨询有限公司
筹备委员会委员：修　龙　张百平　顾勇新　崔　愷　孟建民　庄惟敏　邵韦平
　　　　　　　　沈　迪　李存东　钱　方　薛　明　孙宗列　傅绍辉　高文艳
总策展人：沈　迪
策展团队：高文艳　黄秋平　姜世峰　丁　顺　董　艺　周晓文
支持单位：华东建筑集团股份有限公司
　　　　　北京市建筑设计研究院有限公司
　　　　　中国建筑设计院有限公司
　　　　　中国中元国际工程有限公司

UIA World Architects Congress is held every three years. The 26th UIA Congress will be held in 2017 from Sep. 3 to Sep. 10 in Seoul, South Korean. The congress is hosted by UIA, Federation of Institutes of Korean Architects and Seoul Metropolitan Government, and it is organized by UIA 2017 Seoul Organizing Committee. "Soul of city", UIA 2017 Seoul's main theme, projects strong commitment and determination to acknowledging the importance of having souls dissolved in architecture and cities we live in.

A city, can be considered as a living organism with an exterior and an interior. An exterior of a city can be understood as its apparent impression as entities or its master plan, whereas an interior of a city is seen to be the elements of urban structures such as squares, streets or buildings. Perhaps, the very notion of the theme may be a challenging and perplexing concept to describe; yet the nature of soul can be portrayed as a subject with a mental and immaterial feature.

A soul is an essential element not only in human lives, and so is in cities. To revitalize a city, we must improve the state of its soul and body. Like a soul of a man, a soul of a city can be differentiated from one another, as a soul is fostered by natural process without intentions and a soul cannot be artificially created or manipulated but only "found" in the realities of the terrain on site. Roles of architects, in this context, are indeed significant. They extract and uplift the value of cities, and deliver meaningful and diversified urban environments to life through fusing all elements related to promoting sustainable development.

2017 UIA Congress will bring together architects, engineers and scholars all over the world to Seoul. They will exchange ideas on culture, future, nature and mankind's value, and they will all benefit from discussions by build consensus and seeking common development.

国际建协（UIA）世界建筑师大会每三年举办一次，第 26 届世界建筑师大会于 2017 年 9 月 3 日—10 日在韩国首尔举行。此次会议由国际建筑师协会、韩国建筑师学会联合会、首尔市政府主办，2017 首尔世界建筑师大会组织委员会承办。本次大会的主题是"城市的灵魂（Soul of City）"，意在知觉人类生活家园——城市和建筑——中流淌的灵魂，并认知其重要性。

城市宛如有生命的有机体，由内部和外部组成，城市外部意味着从远处看到的城市外观，或是通过总体规划形成的城市整体印象，城市内部则可以理解为广场、街道、主要建筑物等城市要素。而城市灵魂的概念似乎难以明确地描述和定义，但可以肯定的是，城市灵魂是凌驾于物质之上的非物质因素。

灵魂是人类也是城市最根本的要素。要重振城市，需要改善的不仅是"躯壳"，还要唤醒灵魂。每个城市都有属于自己的灵魂，它通过特定的城市性格、文化传统、情感意志、历史记忆等呈现出来，它并非人为创造，而是在城市发展过程中自然形成，并期待着人们去发现和探寻。此时，建筑师的作用显得至关重要，他们将融合推动城市可持续发展的要素，发掘并提升城市价值，为人们创造极具内涵而丰富多样的城市环境。

2017 UIA 盛会将邀请全球建筑师及技术人员、学者共聚一堂，他们将围绕文化、未来、自然以及人类价值等主题，相互交流、集思广益、凝聚共识、共谋发展。

INTRODUCTION
12

Contents

Urban Renewal
14

Redevelopment
18

Yushu Khamba Arts Center	Library on the Quay	Gymnasium of New Campus of Tianjin University	Huashan Forum and Ecological Plaza
Mogao Grottoes Digital Exhibition Center	Shanghai Chess Institute	Jizhaoying Mosque	Networking Engineering Center, Nanjing Sample Sci-Tech Park
Affiliated High School of Peking University	Happiness Garden Exhibition Hall, Beichuan Earthquake Memorial Park	Creative Valley of South Taizi Lake (Phase I)	China Welfare Institute Pujiang Kindergarten
Humble Administrator's Villa	Customs Clearance Service Center of Chenglingji Free Trade Zone in Hunan	WEI Retreat Tianmu Lake	Shooting Range Hall of the East Asian Games
Yichang Planning Exhibition Hall	Liu Haisu Art Museum	Shenzhen Maritime Sports Base and Marine Navigation Sports School	Shandong Art Gallery
D23 Project, Plot 8 of Hongqiao Business District	Aimer Fashion Factory	2022 The Winter Olympics Plaza	Xi'an North Station of Zhengzhou-Xi'an High-speed Railway
Tianjin TV Station			

Rehabilitation
118

Jixi Museum	Hongqiao International Airport T1 Renovation and GTC Project	Power Station of Art	Nanjing Yu Garden - A Project to Improve the Surrounding Environment
Shougang Museum	Relics Park for the Coal Dock of Xiaguan Power Plant	Xi'an South-Gate Plaza Improvement Project	Historic Block Space Regeneration- Seclusive Jiangnan Boutique Hotel
TENIO Green Design Center	Quadrangle Renovation, Caochang Area, Beijing (Yard 8,19,36,41)	Tianning NO.1 Culture and Creative Industrial Park	Comprehensive Renovation Project of Zhongshan Road (from Jianghan Road to Yiyuan Road)
Hexing Warehouse Renovation, 2010 Shanghai Expo Park, China	Preservation and Reparation Project of the Capital Cinema	Landscape and Related Facilities of The 9th China (Beijing) International Garden Expo Park	Protection and Renewal Plan of Shichahai Neighborhood in Beijing (2013-2030)

Conservation
182

Site Museum of Jinling Grand Bao'en Temple	The Dinosaur Egg Remainder Museum in Qinglong Mountain	The Restoration of Holy Trinity Church	Preservation and Restoration of the Joint Trust Warehouse
Bund 33 # Renovation of the Original British Consulate Building and the Apartment	Protection Plan of Fujian Tulou, the World Cultural Heritage	Fairmount Peace Hotel Renovation and Expansion Project	Preservation and Reparation Project of Shanghai Great World

目录

引言
12

城市更新
14

再开发
18

- 玉树康巴艺术中心
- 敦煌莫高窟数字展示中心
- 北京大学附属中学
- 拙政别墅
- 宜昌规划展览馆
- 虹桥商务区 8 号地块 D23 项目
- 天津电视台
- 码头书屋
- 上海棋院
- 北川抗震纪念园幸福园展览馆工程
- 湖南城陵矶综合保税区通关服务中心
- 刘海粟美术馆
- 爱慕时尚工厂
- 天津大学新校区综合体育馆
- 吉兆营清真寺
- 武汉南太子湖创新谷（一期）
- 天目湖微酒店
- 深圳海上运动基地暨航海运动学校
- 2022 首钢西十冬奥广场
- 华山论坛及生态广场
- 南京三宝科技集团物联网工程中心
- 中福会浦江幼儿园
- 东亚运动会射击馆
- 山东美术馆
- 郑西高铁西安北站

整治改善
118

- 绩溪博物馆
- 首钢博物馆
- 天友绿色设计中心
- 中国 2010 年上海世博会和兴仓库改造
- 虹桥国际机场 T1 航站楼改造及交通中心工程
- 南京下关电厂运煤码头遗址公园
- 北京前门草厂片区四合院改造
- 首都电影院
- 上海当代艺术博物馆
- 西安南门广场综合提升改造项目
- 天宁一号文化科技创新园
- 第九届中国（北京）国际园林博览会园区绿化景观及相关设施建设项目
- 南京愚园地块保护与整治改善
- 历史街区空间再生——隐居江南精品酒店
- 中山大道综合改造工程（江汉路至一元路）
- 什刹海街区保护与更新发展规划（2013—2030）

保护
182

- 金陵大报恩寺遗址博物馆
- 上海外滩源 33# 原英国领事馆及官邸历史建筑保护及再利用工程
- 青龙山恐龙蛋遗址博物馆
- 世界文化遗产福建土楼保护规划总纲
- 基督教圣三一堂修缮工程
- 和平饭店修缮与整治工程
- 四行仓库修缮工程
- 上海大世界修缮工程

9

	Preservation Project of Prince Gong Mansion	Restoration of the Building of Shanghai Kunju Opera Troupe		
Rural Reconstruction 222				
Protection and Regeneration of Traditional Villages 226	Homeland of Mosuo, The Project of Protecting Mosuo Habitation	Shaxi Rehabilitation Project	The Practice of Protection and Utilization of Traditional Villages in Yangjiatang Village, Songyang County	Planning of Xiedian Traditional Village Protection and Regeneration
Construction and Renewal of Beautiful Villages and Towns 242	Incremental Update Design of Gunan Street Historical Culture Districts in Yixing			
Rural Architecture 246	Zhudian Hoffmam Kiln Culture Center	Macha Village Community Center	Contemporary Collective Living: New Forms of Affordable Housing for Relocalized Farmers in Hangzhou	Renovation of Countryside Buildings of Xibang Village
	THE-Studio (Tsinghua Eco Studio)	Renovation of Wencun Village	Hani Nationality Mushroom House Retrofit Experiment in Azheke Village, Yuanyang County	A Rural Shop at Huashu Village, Nanjing
	Oriental Rural Life Pavilion, Huishan, Wuxi	Comprehensive Building and Academic Center of FAST Engineering Observation Base	The Yinlu Tea House of Lianhuadang, Yixing	Lounge Bridge Renovation
	Wooden Bridge	Renovation Project of Grandma's Yard in Huangshandian Village	Mendeng Village Community Center	
Chinese Architects 306				
Appendix 309	Index	Introduction of Chief Curator	Sponser Introduction	Organizer Introduction

	恭王府府邸文物保护修缮工程	上海昆剧团大楼修缮改造工程		

乡村建设
222

传统村落
保护与再生

226

| | 摩梭家园——摩梭人聚居地保护 | 沙溪复兴工程 | 松阳县杨家堂村传统村落保护与发展实践 | 谢店村传统村落保护与再生规划设计 |

美丽村镇
建设与更新

242

| | 宜兴市古南街历史文化街区渐进式更新设计 | | | |

田园建筑

246

	祝甸砖窑文化馆	马岙村村民活动中心	乡村低收入住宅——杭州富阳东梓关回迁安置农居	西浜村农房改造工程
	清控人居科技示范楼	文村村新建民居	元阳县阿者科村哈尼族蘑菇房改造实验	南京桦墅村口乡村铺子
	无锡惠山东方田园生活馆	FAST 工程观测基地综合楼和学术交流支撑平台及附属设施工程	隐庐莲舍：宜兴莲花荡农场茶室	驿道廊桥改造
	木桥听雨	黄山店姥姥家民宿改造项目	门等村村民活动中心	

中国建筑师

306

附录

309

| | 工程列表 | 总策展人介绍 | 主办单位介绍 | 承办单位介绍 |

INTRODUCTION

The theme of the Architecture Society of China of 2017 UIA congress is "Fusion and Harmony". China has been experiencing a rapid growth over the past four decades, since the implementation of the reform and opening-up policy. Its cityscape and urban culture have undergone fundamental changes, as massive construction booms, buildings are diversified, urban population surges and urban life style changes. Today, Chinese economy has entered a new phase that is different from the high-speed growth pattern exhibited in the past. It is a new trend that features more sustainable growth with higher efficiency and lower cost. In rural and urban construction, China faces new challenges to maintain sustainable development. Chinese architects are ready for the new stage in which they seek to achieve harmonious fusion and discover the enduring truths in architecture, regardless of prevailing world trends. "Fusion" is a core word in this stage. It is a way of thinking that is applicable to dealing with big things like relations of countries and small things like spacial relations of buildings.

Likewise, the fusion thought can be applied to urban-rural and architectural fields when dealing with relations of environment and human, history and contemporary era, and cities and villages. Chinese architects begin to seek a road of harmonious development. They rethink former growth patterns, the pattern of "little investment but quick results" and the pattern of "flashy and strange new things with little substance" that have brought conflicts and incongruity.

"Fusion and harmony" is a dynamic process, starting from course of fusing to outcome of integrating as one. Now China has been experiencing this stage of development. "Fusion and harmony" is a modern interpretation of traditional Chinese wisdom, both spiritually and technically, though the thought may not be limited to Chinese architects but also be shared by architects worldwide.

Our display includes three sub-themes, namely, urban renewal, rural construction, and Chinese architects. The first two themes respectively introduce Chinese urban renewal that emphasizes harmonious development, and vernacular architecture that follows Beautiful Villages strategy instead of former urbanization. For each of the two themes, we have selected representative works that were done by Chinese architects and have been built up and in recent 10 years. The works are of different types, are located in different regions, and are created by different design entities. The last sub-theme focuses on Chinese architects, the creators of buildings. It covers their lives, work and social responsibilities, and tries to depict the fusion of their cross-disciplinary versatility.

The book has collected all the works of urban renewal and rural construction that are displayed in the Congress, 71 in total. The video about Chinese architects is burned into CDs and they are presented with the books. It is to record the exhibition and let more people know the development and the transformation of Chinese architecture.

引言

本次中国建筑学会展览的主题为"融—合之间"。历经近四十年的改革开放，中国经历了快速发展的重要历史时期，几十年间，城市和建筑大规模崛起，人口急剧增加，城市生活方式改变，建筑种类不断更新，城乡面貌和城市文化变化显著。如今，中国经济从长期高速发展进入增质提效、转型调整的"新常态"；中国城乡建设面临着从高歌猛进转向可持续发展的新挑战；中国的建筑创作经历了从纷扰迷茫进入回归本源、寻求融合的新阶段。"融合"成为这一阶段的核心词，也成为中国人思考当下大到国与国关系，小到空间关系时采取的思维模式。

这一思维模式在城乡、建筑等广阔的图景中得以体现，在面对环境与人类，历史与当代，城市与乡村等关系时，中国建筑师开始反思以往的发展模式，反思"短平快"或"新奇异"带来的对立与冲突，转而探索相融与和谐的发展之路。

"融—合之间"是一个动态过程，从融开始，追求最终的合，而当下中国正处在这一历史发展阶段之中。"融—合之间"也是中国古代智慧在当代的运用，从精神层面映射到具体技术层面，反映当下中国建筑师乃至全球建筑师的思考。

本次展览共分为城市更新、乡村建设和中国建筑师三个主题板块，前两个板块分别展示了追求相融与和谐的中国城市更新发展之路，以及从城镇化到美丽乡村的中国乡村本土建筑之路，从多角度呈现了不同地域、不同类型、不同设计主体的建筑活动，在每个主题选择了近 10 年建成的有代表性的中国建筑师的原创作品做深入介绍；第三个板块将展示的对象从建筑本身转向建筑的设计者——中国建筑师们，通过他们的工作、生活、社会责任等方面映射中国建筑师多元融合的发展状态。

本书汇集了本次展览中城市更新和乡村建设板块的所有参展作品，共计 71 个，并将中国建筑师板块的视频资料刻录成光盘随书呈送，以此记录本次展览的主要内容，并让更多读者了解中国建筑的发展与转型。

Urban Renewal 城市更新

Urban renewal is an age-old topic. Cities, like living organism, have been experiencing metamorphosis all the time, and from very beginning are accompanied with urban renewal all the time.

Four decades of high speed economic development in China have greatly promoted its city's growth, providing much better transportation, living and business environment. Meanwhile it also brings problems like increasing shortage in land resources, breakage of city's historical context, and loss of features in regional culture. People begin to rethink the pattern of leapfrog development, and pay more attention to renew the stock instead of constructing new ones. Renovating the existing building stock, intensively utilizing land resources, and uplifting environment are the key aspects of the city's future development. The extensive growth mode of large-scaled demolishing old buildings and constructing new ones will gradually being replaced by the new trend that highlights heritage protection, cultural inheritance and sustainable growth.

With this background, Chinese architects attach great importance to the integration of comprehensive development through urban redevelopment, rehabilitation and conservation. They are dedicated to inaugurate ways to renew the city and contribute to a harmonious society.

Redevelopment

Redevelopment plays an important role in urban renewal in China. Due to structural adjustment and industrial upgrading, many built-up areas including old industrial fields, docklands, warehouses and residences nowadays can no longer keep up the pace with social economic development. In process of urban renewal, these areas can be redeveloped so that they can adapt to the requirements of the new situation through changing either their land properties or their functional programs. The goal is to optimize urban functions, industrial structure and living environment.

Rehabilitation

In process of urban development and construction, it is important to upgrade aging facilities, improve battered buildings and restore deteriorated land. The old downtown areas and old buildings can be revived and the overall environment in old areas can be uplifted if we make adaptive reuse, strengthen structures to extend their service lives, and update their functions.
We are encouraged to protect the buildings and city blocks that bear memories, have values and worthy of being reused. The goal is to restore and promote cultural inheritance, keep the city's memories and seek intensive development through comprehensive renovation, functional conversion and environment uplifting.

Conservation

Historical context bears city spirit and cultural accumulation. It is the soul and the root of the city, and Chinese architects have widely recognized principle of conservation. They revive historical buildings through repair, reusing and through their own practices. They are trying to knit up the fragments of history context, and extend the city's history far and wide.

城市更新是个古老的话题，从城市产生一直到今天，城市更新伴随城市发展，如影随形，城市仿佛有机的生命体，不断地寻求蜕变。

经过近40年的大发展，中国城市的交通、居住、商业等物质环境得到了极大的改善。但同时也导致了土地资源的日益短缺以及城市历史、地域文化特色的不断消失，这使得中国当下的城市建设从跨越式发展进入反思期，从关注增量进入到存量更新的发展阶段。存量建筑的改造，土地的集约化利用和环境质量的提升将成为未来城市发展的重点，而"大拆大建"、"推倒重来"的粗放式城市更新方式，也逐渐向注重遗产保护和文化传承的有机更新方向转化。

在这种大背景下，中国建筑师开始关注城市建设的综合性和整体性，通过对城市的再开发、整治改善及保护等不同模式的探索，探讨城市更新的路径和方法，提高城市环境品质，让城市更温暖和谐。

再开发

再开发是中国目前城市更新发展的主要模式。由于功能和产业结构的二次调整，城市建成区，包括旧工业区、码头区、仓库区、住宅区等城市物质空间的使用功能不再适应当前社会经济发展需求，城市更新通过土地性质及建筑功能的改变等活动，对城市土地进行再开发来满足新的城市功能需要，从而完善城市功能，优化产业结构，改善人居环境。

整治改善

对设施老化、建筑破损、环境不佳的地区和建筑进行整治改善逐渐成为我国城市发展和建设过程当中的重要手段。通过对区域或建筑进行不同程度的适应性改造，延长建筑的使用寿命，保证建筑的安全使用，赋予建筑全新的功能，全面改善旧城地区的生产、生活和居住环境，使旧城区域或老建筑获得再生。

对具备改造再利用价值和条件、承载城市记忆的建筑和街区，通过有效保护、综合改造、功能置换和环境提升，推动城市文化的回复和传承，保留城市记忆，实现城市内涵式发展。

保护利用

历史文脉是一座城市精神与文化的积淀与凝结，是一座城市的灵魂与根源所在。"保护"已成为中国建筑师的集体共识，建筑师通过积极实践，保护修缮和再利用的方式来延续历史建筑的活力，使历史建筑获得重生，织补碎片化的历史文化肌理，加强城市历史的纵深感。

Yushu Khamba Arts Center 玉树康巴艺术中心

Location: Yushu, Qinghai Province, China
Architect (Studio): Cui Kai, Guan Fei, Zeng Rui, Dong Yuanzheng, Gao Fan, China Architecture Design Group
Design: 2011
Construction: 2014
Site Area: 24,563m²
Floor Area: 20,610m²

地点：中国青海省玉树藏族自治州
建筑师（事务所）：中国建筑设计院有限公司：崔愷、关飞、曾瑞、董元铮、高凡
设计时间：2011
建造时间：2014
用地面积：24563m²
建筑面积：20610m²

Khamba Arts Center is located at Jiegu Town, the most afflicted city during Yushu Earthquake. The complex includes theatres, theatrical troupe offices, a cultural hall and a library. To show respect to the traditional city texture, the layout is loose fit to the axes of the neighbouring streets and square, especially the Ta'er Lamasery. Architects limit the scale of the building but increase the use efficiency of different areas. Courtyards and terraces are revealed adequately to trace the atmosphere of the traditional Tibet building spaces. Bright colours economically enrich the visual effects of the building and arouse the memory of Tibet patterns.

玉树康巴艺术中心是为遭受玉树地震后的结古镇所作的重建工程，汇集了原玉树藏族自治州的剧场、剧团、文化馆和图书馆等多种功能。设计从尊重城市文脉的角度出发，总体布局自由松散但错落有致，强调其与塔尔寺、唐蕃古道商业街、格萨尔广场等周边城市元素的对位呼应。从建筑的密度上与传统城市肌理相吻合，步行街道的尺度也尽力与唐蕃古道商业街相协调。本着经济、环保的原则，设计将州剧团的辅助功能区与剧院的后台区合并，将排练厅与室外演艺功能合并，尽可能控制规模，并强化各功能区的通用性。平面布局力图通过再现院落空间的组合体现传统藏式建筑的空间精神，并尝试在建筑布局上体现台地特征，建筑在体量上也逐层递减。基于低造价的考虑，建筑通过沿袭传统藏式建筑丰富的色彩形成丰富的视觉效果。

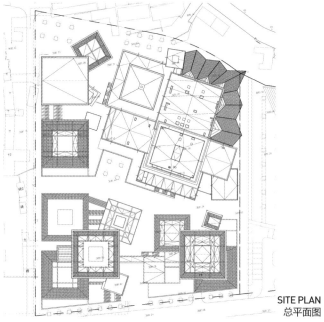

SITE PLAN
总平面图

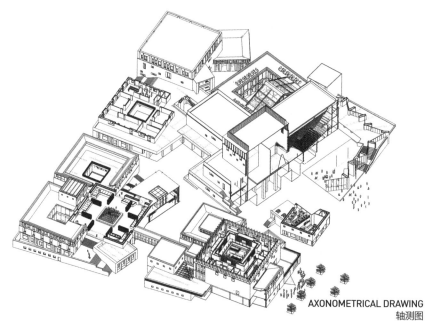

AXONOMETRICAL DRAWING
轴测图

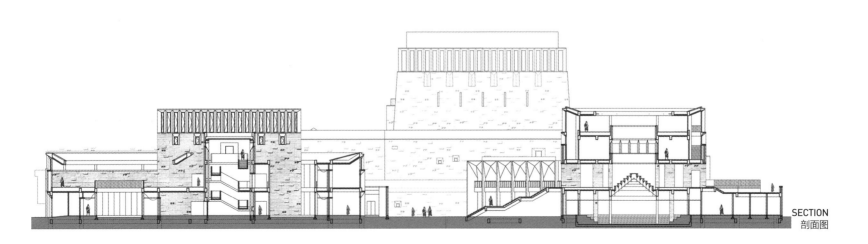

SECTION
剖面图

Library on the Quay 码头书屋

Location: Riverside Park, Tongling, Anhui Province, China
Architect (Studio): Architects & Engineers Co., Ltd of Southeast University
Design: 2015.02-2015.05
Construction: 2015
Site Area: 853m²
Floor Area: 517m²

地点：中国安徽省铜陵市，滨江公园
建筑师（事务所）：东南大学建筑设计研究院有限公司
设计时间：2015.02-2015.05
建造时间：2015
用地面积：853m²
建筑面积：517m²

This project is a library built on an abandoned quay by the riverside. The library is an important part of the riverside park that was converted from an industrial wharf. The purpose of the library is not only to provide a place to read books with fantastic outdoor sceneries, but also to preserve the historical memory of the place. The quay, 40 meters long and 14 meters wide, is constructed of rubble and concrete. The main structure of library is 32 x 14 meters steel structure which is supported by 6 reinforced concrete columns. The 1st floor of the library is elevated to expose quay space. The 2nd floor is the reading room. The top floor is a platform that could see the whole river views. The rough rubble, solid steel and soft bamboo reflect the Chinese's natural view.

The inner space of the library is surrounded by bookshelf walls in three sides. The desks are set close to the bookshelf walls. People can see outside river through the window. The stair type reading area is cascading towards the river. The ceiling of the reading room is decorated with a bamboo device, an upside-down bamboo views. It wings gently in the breeze, an artistic aesthetic commonly presented in Chinese culture.

本项目是利用江边的一座废弃码头，建造一个书屋。书屋是工业码头改建而成的滨江公园的重要组成部分，其目的是为市民提供阅读书籍和观景的场所，同时也保留人们对场所的历史记忆。码头采用毛石和混凝土砌筑，长约40m，宽约14m，设计中32m×14m的书屋主体钢结构框架被6根钢筋混凝土柱子支撑在码头上，大尺度的结构出挑既形成了强烈的空间张力，也切实降低了对码头的破坏程度，同时避免建筑受江水袭扰。架空建筑形体的底层为原有码头空间，中层是书屋的阅览空间，顶层是观江平台。粗糙的毛石、硬朗的钢、温暖的竹三者之间的平衡关系表达出当下中国人的自然观。

书屋室内空间是沿三面环绕的书架墙展开，书桌紧靠书架布置，靠窗的落地玻璃满足观江的需要，书架墙的内部围合出一个由阅览室向码头表面跌落的"阶梯阅读区"，该空间的吊顶是一个竹装置——"倒置竹山水"。竹装置在风中轻轻晃动，仿佛跃然眼前的水墨山水，使人产生奇妙的感受。这就是中国人常说的，"宁可食无肉，不可居无竹"的意境。

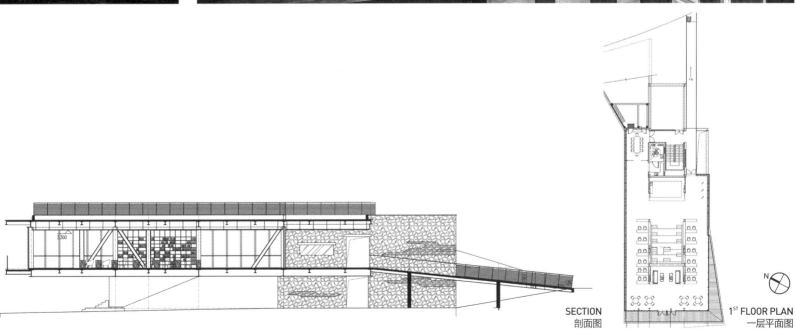

SECTION
剖面图

1ST FLOOR PLAN
一层平面图

25

Gymnasium of New Campus of Tianjin University

天津大学新校区综合体育馆

Location: Jinnan District, Tianjin, China
Architect (Studio): Atelier Li Xinggang, China Architecture Design Group
Design: 2011.02-2013.08
Construction: 2015.11
Site Area: 33,950m²
Floor Area: 18,798m²

地点：中国天津市津南区
建筑师（事务所）：中国建筑设计院有限公司李兴钢建筑工作室
设计时间：2011.02-2013.08
建造时间：2015.11
用地面积：33,950m²
建筑面积：18,798m²

The Gymnasium of New Campus of Tianjin University is located to the north side of the existing campus. The public space of gymnasium and natatorium is connected by a huge slight-arched corridor. The various kinds of indoor sport venues are compactly arranged according to sizes, headroom clearances and usages, and are integrated by some overlapped or tandem linear public spaces. A series of ruled surface, barrel-vaulted and coniform steel concrete structures enable large-span spaces and clerestory windows. The design exposes texture of timber formwork concrete inside, and creates a solemn and dynamic architecture profile outside. The extreme sport area spreads from the irregular exterior steps and terraces to the gradually changed ruled surface of the waved roof of public hall. The design of 140-meter-long indoor runway highlights natural daylight, and the windows emphasize the roof curves and bring in outdoor sceneries. This is how the design harmoniously links indoor and outdoor spaces, and the ground level and rooftop, and makes it an "integrated sport complex".

天津大学新校区综合体育馆，位于校前区北侧，包含体育馆和游泳馆两大部分，以一条跨街的大型缓拱形廊桥将两者的公共空间串通。各类室内运动场地依其对平面尺寸、净空高度及使用方式的不同要求紧凑排列，并以线性公共空间叠加、串联为一个整体。一系列直纹曲面、筒拱及锥形曲面的钢筋混凝土结构，带来大跨度空间和高侧窗采光，在内明露木模混凝土筑造肌理，在外形成沉静而多变的建筑轮廓。极限运动区通过不规则铺展的室外台阶看台，可以一直延伸到公共大厅波浪形渐变的直纹曲面形屋面。东侧长达140m的室内跑道，为大厅带来凸显屋面形状的自然光线和向远处延伸的外部景观。如此成为一个室内与室外、地面与屋面联为一体的"全运动综合体"。

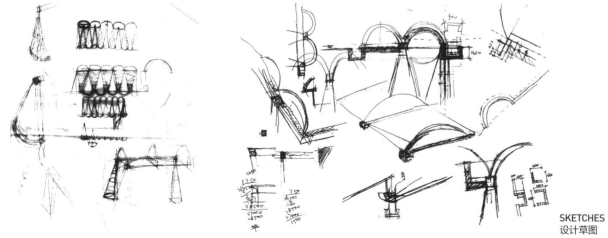

SKETCHES
设计草图

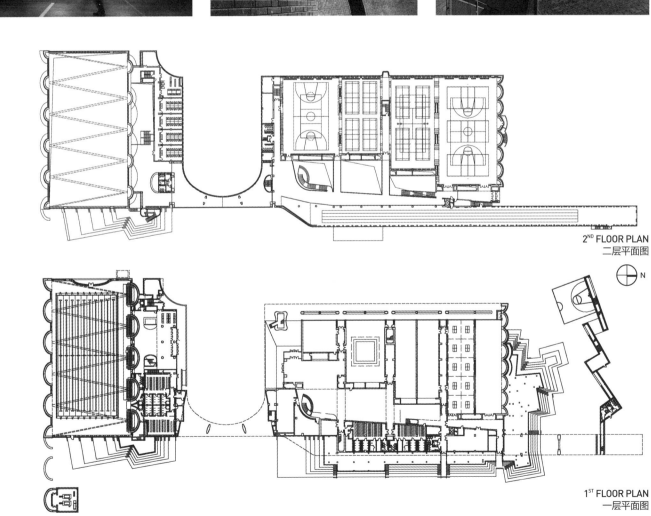

2ND FLOOR PLAN
二层平面图

1ST FLOOR PLAN
一层平面图

Huashan Forum and Ecological Plaza

华山论坛及生态广场

```
Location: Huayin, Shaanxi Province, China
Architect (Studio): Architectural Design and Research
Institute of Tsinghua University Co., Ltd.
Design: 2008.08-2009.11
Construction: 2011.04
Site Area: 408,008m²
Floor Area: 8,867m²

地点：中国陕西省华阴市
建筑师（事务所）：清华大学建筑设计研究院有限公司
设计时间：2008.08-2009.11
建造时间：2011.04
用地面积：408008m²
建筑面积：8867m²
```

Huashan Forum and Ecological Plaza project site is located to the south of 310 National Highway of Huayin, Shaanxi, 5 kilometers from the city south and on the north foot of Mount Hua, a famous national scenic area. The site is adjacent to Mount Hua in the south and Yingbin Avenue in the north, overlooking the train station in the north tip of the Avenue. On the east and west sides of the site are fields and scattered farmhouses, with unobstructed views. The renowned cultural relic Xiyue Temple is 4 kilometers on east from the north side of the site, which in the planning will be connected to the project through Gubai Pedestrian Street. The southeast site is higher than the northwest corner. In general, the north part of the site slopes gently and the south part has a bigger height difference.

Considering most tourists are coming to visit Mount Hua and that a Tourist Center is only there to provide necessary services for these visitors, the volume is preferred small rather than big, hidden rather than conspicuous. The buildings lie prostrate on the foot of Mount Hua and are integrated into the natural environment. The main function of the Tourist Center is to serve visitors. It is a small and comprehensive building that integrates functions including tourist distribution, information service, guide service, touristic shopping, catering and office management. It also bears essential cultural implications as it is located in Guanzhong (mid Shaanxi province), which is rich in traditional Chinese culture. The guiding principles for design are: first, it is a multi-functional, mixed-used building with high standards and certain cultural implications; second, its planning and architectural design are integrated and symbiotic with the surrounding environment; third, it is people oriented and takes advantage of a good natural environment in precondition of protecting natural landscape of Mount Hua. For these ends, unique planning and designing concepts are introduced to insure unification of approaches, technics and arts.

华山论坛及生态广场项目用地位于陕西省华阴市城南5km处，310国道以南，著名的国家级风景名胜区华山的北麓。用地南依华山，北侧正对华阴市迎宾大道，与迎宾大道北端的火车站遥遥相望。东、西两侧现为农田和部分散居的农户，四周视野开阔。北侧偏东约4km处为著名的文物古迹西岳庙，并在规划上通过古柏步行街与西岳庙相连。整个场地东南高，西北低。其中北侧地势较为平缓，南侧地势落差较大。

考虑到大量游客来此旅游都是为了观仰华山，而游客中心仅仅是为其观山提供必要的服务。因此，在设计立意上，整个建筑体现了宜小不宜大、宜藏不宜露的原则。建筑匍匐在华山脚下，融于用地的自然环境之中。此外，华山游客中心承担着服务游客的主要职能，是高水准的集游客集散、咨询服务、导游服务、旅游购物、餐饮及配套办公管理等功能于一体的综合性小型建筑；并且是与关中地区传统文化地位相称的具有文化内涵的重要建筑。它的建设遵循这样几个设计原则：首先，它是具有高品位和一定文化内涵的综合性建筑；其次，在规划及建筑设计上与周边环境相融共生；最后，以人为本，依托良好的自然环境，在保护华山自然风貌的前提下，为游客提供便利服务。在该项目设计中引入了独特的规划及设计理念，从而达到设计方法、技术手段和建筑艺术的统一。

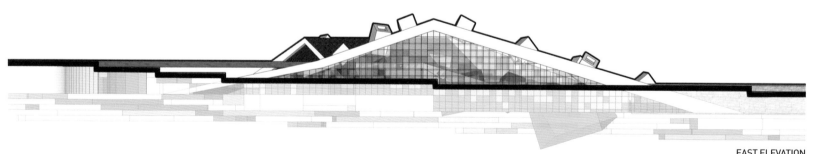

EAST ELEVATION
东立面图

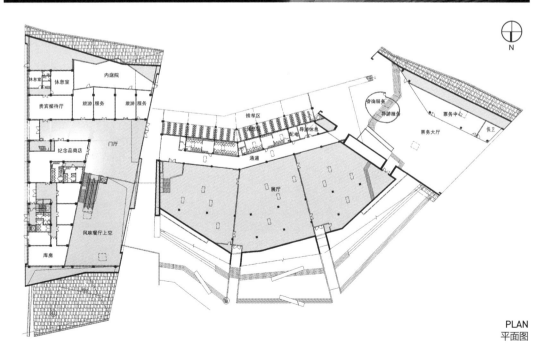

PLAN
平面图

Mogao Grottoes Digital Exhibition Center

敦煌莫高窟数字展示中心

Location: Dunhuang, Gansu Province, China
Architect (Studio): Cui Kai, Wu Bin, Feng Jun, China Architecture Design Group
Design: 2008
Construction: 2014
Site Area: 40,000 m²
Floor Area: 10,440m²

地点：中国甘肃省敦煌市
建筑师（事务所）：中国建筑设计院有限公司本土设计研究中心：崔愷、吴斌、冯君
设计时间：2008
建造时间：2014
占地面积：40000 m²
建筑面积：10440m²

Mogao Grottoes Digital Exhibition Center is located 15 kilometers away from the Mogao Grottoes. To protect the precious cultural legacy, most of the exhibition and tourist service functions are set here. We designed the building as a flowing volume with sand-like surface and dune shape, which help to harmonize it with the its environment. The double ventilating roof can reduce the solar heat effectively, and the underground ventilating tube can cool down the air flow and reduce the cooling load.

莫高窟被誉为"东方艺术宝库"，但庞大数量的游客对遗产的保护和管理造成很大困扰。这座建于绿洲和戈壁之间的莫高窟数字展示中心，即为缓解景区的保护压力而建，集合了游客接待、数字影院、球幕影院、多媒体展示、餐饮等功能。

设计伊始，最初的感动来自对大自然的敬畏和对古代工匠精美艺术的敬佩。这座建筑，应该是大漠戈壁中的一座小沙丘，造型既如同流沙，如同雅丹地貌中巨舰般的岩体，又类似矗立在沙漠中的汉长城，莫高窟壁画中飞天飘逸的彩带，充满着强烈的流动感。若干条自由曲面的形体相互交错，婉转起伏，巨大的尺度和体量将沙漠地景建筑的特征表达得淋漓尽致。

充满动感的形态特征从室外延续到室内，所有的公共功能区均为开放空间，顺应外部形态的变化，室内空间的高度也随之变化。结构支撑体的形态用"墙"的概念，将不同功能、不同高度的空间进行划分，界面清晰明确。

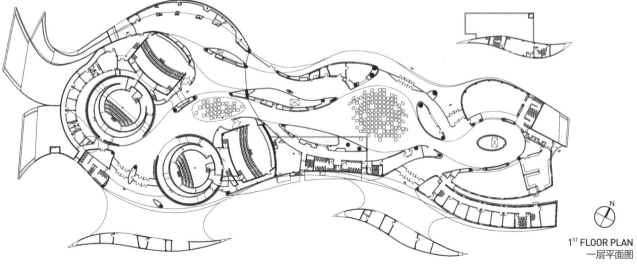

1ST FLOOR PLAN
一层平面图

Shanghai Chess Institute

上海棋院

Location: Shanghai, China
Architect (Studio): Zeng Qun, Tongji Architectural Design (Group) Co.,Ltd.
Design: 2012.08-2013.02
Construction: 2013-2016
Site Area: 6,002m²
Floor Area: 12,424m²

地点：中国上海市
建筑师(事务所)：同济大学建筑设计研究院(集团)有限公司；曾群
设计时间：2012.08-2013.02
建造时间：2013-2016
用地面积：6002m²
建筑面积：12424m²

Noisy / Peaceful

The project is situated at the bustling West Nanjing Road in Shanghai. The essence of Chinese traditional architecture is presented by including gardens and recreating traditional space with modern approach. Presenting itself quietly and peacefully at the busy West Nanjing Road with heavy commercial circulation, it segregates itself from the surrounding buildings and highlights its own cultural image.

Open / Introverted

The pocket-like site is long and narrow, about 140 meters long from north to south and 40 meters wide from east to west at the narrowest point. The outerspace close to commercial street is open, while the inner space of "pocket" is introverted.

In width direction, the staggered spaces of void and solid enable a capacious feeling inside a restricted area. Courtyards are surrounded by walls, and the walls are infiltrated by courtyards. The "checkerboard" side walls, like a light sieve, gently blur the inside and the outside. The programs follow the feature of the site, and are set along the long direction. As people go inside, the characters of space transfer from dynamic to serenity, from extroversion to introversion. Sequence of programs is arranged correspondingly, from open lobby, to public game hall and finally to the exhibition rooms.

喧嚣 / 静谧

上海市繁华的南京西路段，一幢体育文化建筑如何自然地介入这个喧嚣的都市商业氛围中，是这个设计的重点。

设计试图通过院与墙的结合融合中国传统建筑的精髓，以现代的手法体现传统空间，建筑整体形态完整统一，庭院的运用使得建筑整体充满了中国意味。以安静祥和的姿态出现在充满商业氛围的南京西路，与周边建筑形成强烈的对比和反差，从而突出建筑的文化形象。

开放 / 内敛

项目基地为南北向狭长"口袋"地块，南北长约140m，东西最窄处约40m。沿街道处空间是对城市开放的，越往里走氛围愈加收紧，空间逐渐内敛。

在面宽方向，设计中将室内和室外的虚实空间交错布局，以墙围院，以院破墙，从而在狭小的用地内争取外部空间。变幻的"棋盘"侧墙像是一个光筛，自然过渡了建筑内外，顺应着功能而有机变化。

在纵深方向，功能布局顺应了基地特质，从外往里，空间从动走向静，从开放走向内敛，对应的功能由开放门厅过渡到公开比赛大厅，再到内向的展厅。

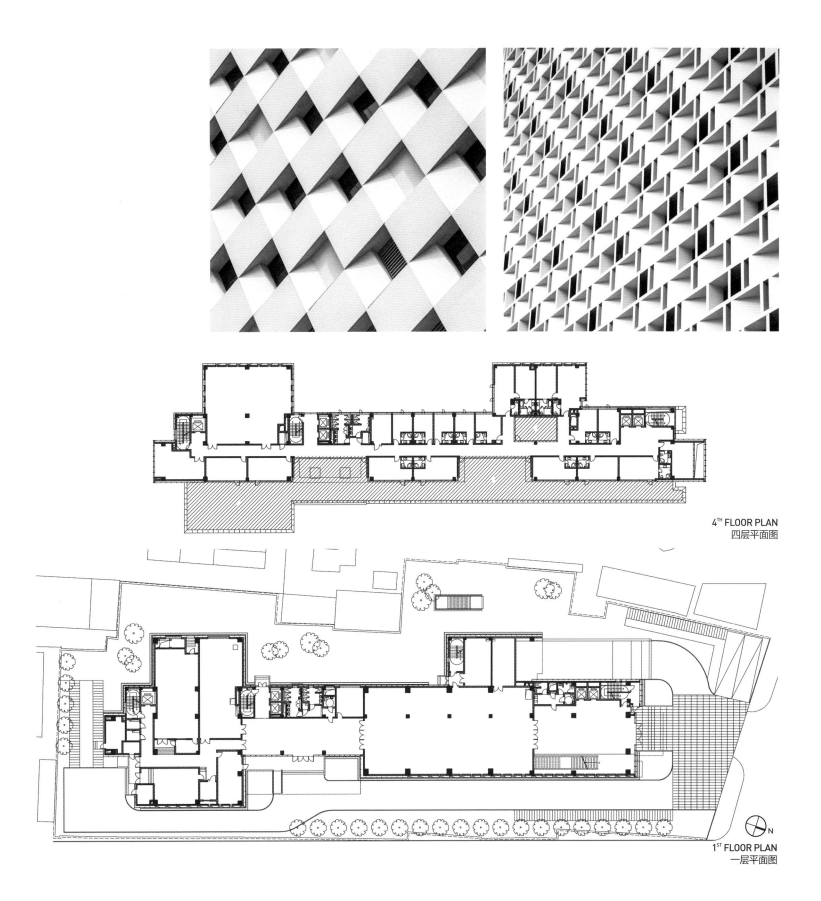

4TH FLOOR PLAN
四层平面图

1ST FLOOR PLAN
一层平面图

Jizhaoying Mosque 吉兆营清真寺

Location: Nanjing, Jiangsu Province, China
Architect (Studio): Architects & Engineers Co., Ltd of Southeast University
Design: 2009.02-2010.01
Construction: 2011.12-2014.01
Site Area: 661m²
Floor Area: 1,307m²

地点：中国江苏省南京市
建筑师（事务所）：东南大学建筑设计研究院有限公司
设计时间：2009.02-2010.01
建造时间：2011.12-2014.01
用地面积：661m²
建筑面积：1307m²

The old Jizhaoying Mosque, the only mosque that remains in the north of Nanjing, is a mosque of traditional courtyard-style built for Chinese Hui people. After vicissitudes of history, the original buildings in the mosque could no longer support daily usage. With background of urban renewal, the reconstruction of Jizhaoying Mosque has significantly improved the venue where local Muslims engage in religious ceremonies and social activities. At the same time, the project uplifts urban environment and fits with the trend of urban development.

At the level of urban planning, the overall layout of the new mosque makes a breakthrough in common setback and density requirements. The design carefully follows site conditions and faithfully retains original boundaries of the old mosque except for necessary road setbacks. While reconciling relations with neighboring environment, the design successfully resolves the conflict of increased demand for usage and decreased site area, integrating "fragments" in urban space.

At the level of architectural design, the design of the new mosque hinges on reformation of space and innovative reuse of old construction materials. Firstly, the design aggregates the courtyard space - which is horizontally organized in traditional Chinese mosques - in a vertical direction so as to organize different functional spaces. Secondly, utilizing trees that exist at the site, the architects transform the prayer space from an inward, closed hall to a space that connects the interior and the exterior, creating a new experience for prayers. Another highlight is reorganization of construction materials. The design appropriately preserves and utilizes remaining objects and materials from the demolished old mosque, sustaining memories of history with a localized design language. Overall, the design is an organic integration of Islamic culture with local cultures in the region to the south of the Yangtze River.

吉兆营老清真寺是南京城北仅存的一座清真寺，也是一座中国回族传统院落式清真寺。原寺历经沧桑，已不能满足正常使用要求。在城市更新背景下，吉兆营清真寺翻建工程很大程度改善当地穆斯林的宗教活动与社会交往的场所，同时适应了城市发展，改善了城市环境。

在城市规划层面，新寺总体布局突破了关于退让与密度的通常规定，采取因地制宜的策略，除必要的退让道路红线，谨守旧寺原有边界。在调适邻里的同时，成功化解了功能需求增长与用地减少的矛盾，整合了城市空间的"碎片"。

在建筑设计层面，新寺采用了空间重塑和旧物新用的设计方法。首先，将传统清真寺水平组织的院落进行竖向叠加，以组织各功能空间。其次，礼拜空间利用原有树木，由内向封闭转变为内外融合，从而提供新的礼拜空间体验。最后，进行建造材料重组，新寺恰当地保护和合理地利用老建筑遗存物件和材料，以本土化的设计语言延续了历史的记忆，展现了伊斯兰文化与中国江南地域文化的有机结合。

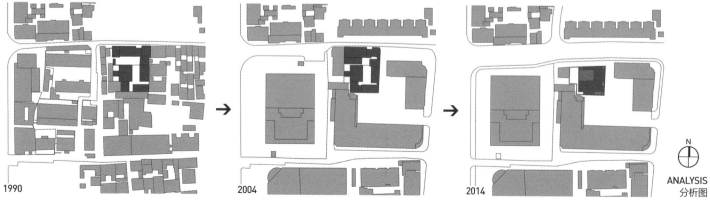

1990 → 2004 → 2014

ANALYSIS
分析图

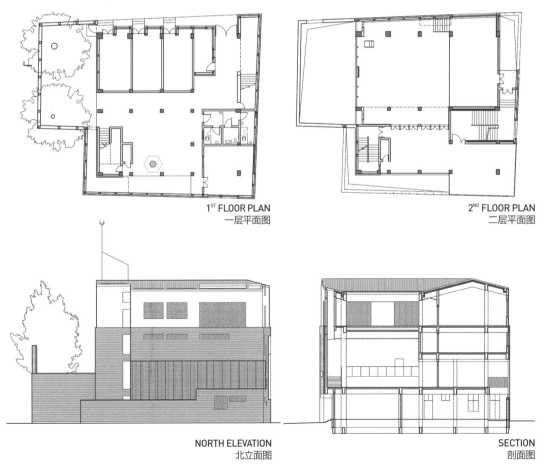

1ST FLOOR PLAN
一层平面图

2ND FLOOR PLAN
二层平面图

NORTH ELEVATION
北立面图

SECTION
剖面图

Networking Engineering Center, Nanjing Sample Sci-Tech Park

南京三宝科技集团
物联网工程中心

Location: Nanjing, Jiangsu Province, China
Architect (Studio): Zhang Tong, Southeast University School of architecture; Architects & Engineers Co., Ltd. of Southeast University
Design: 2010.07-2013.11
Construction: 2012.04-2014.07
Site Area: 24,600m²
Floor Area: 21,505m²

地点：中国江苏省南京市
建筑师（事务所）：东南大学建筑学院、东南大学建筑设计研究院有限公司：张彤
设计时间：2010.07-2013.11
建造时间：2012.04-2014.07
用地面积：24600m²
建筑面积：21505m²

Space Refabrication in the Era of Post-urbanization
The project is situated in the most representative fragmentized urban fabric during the recent 20 years of Chinese urbanization. To remedy the general fragmented environment of Ma Qun Industrial Park, at the eastern foot of Zijin Mountain, this project practises the design strategy of "space refabrication", by reorganizing interrupted and disordered historical and geographical traces, searching for potential opportunities, and reconstructing the continuity and identity of urban texture.

Space Refabrication: Replenishing, Revealing and Promoting
The volume starts from the neglected north-eastern quadrant of existing complex, through the L-shape transition, closing up the central plaza, which used to lead people from north entrance to the sci-tech park. By locating the main building and entrance plaza at the end of the east-westward road, another underlying axis of the area is revealed, and the main entrance becomes an ending of the 300-meter long spatial sequence. The boundary of the park's external environment, the centre, axis, and transition are, thus, specified, and the spatial framework is built up.

Materiality: Another Longitude-latitude System of Space Refabrication
For the refabricated spatial system of Sample Sci-Tech Park, the design uses four materials on the façades, namely, customized ceramic panels (compound windows), glass curtain walls in framed grids, shading surfaces of metal mesh and shading façades of serrated panels. They are organizational elements of Space Refabrication, overlapped as an interweaving structure and bestowing qualities.

后城市化时代的空间织补
项目所在场地是中国20年来城市化过程形成的最为典型的碎片化肌理。作为对紫金山东麓马群工业园区总体环境的弥补，该项目实践的是一种后锋性的"空间织补"策略，在现状似有似无的线迹之间，在被打断和错乱的历史与地理信息中，寻找潜在的机会，重新建构城市肌理的连续性和识别感。

空间织补：补缺、揭示与提升
体形的生成从现状建筑群被遗漏的东北象限开始，通过L形转折，闭合了从北入口进入园区的中心空间。通过将主楼与入口广场置于东西向道路的终端，揭示了园区中潜在的另一条轴线，建筑的主入口成为300m长空间序列的礼仪性终端。至此，园区外部环境的边界与中心、轴线与转接得以明确，空间的框架建立起来。

材质：空间织补的另一种经纬
在三宝园区织补重构的空间系统中，立面的四种材质，定制陶板（复合窗）墙面、框栅玻璃幕墙、金属网板遮阳表层与锯齿板遮阳立面，成为空间织补的组织性元素，叠合成为一种经纬交织的结构，在可感知的物的层面，赋予空间织补以质感。

1ST FLOOR PLAN
一层平面图

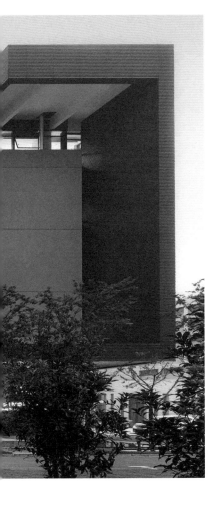

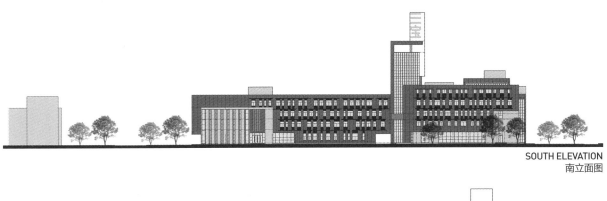

SOUTH ELEVATION
南立面图

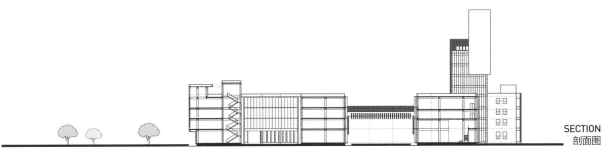

SECTION
剖面图

Affiliated High School of Peking University

北京大学附属中学

Location: Beijing, China
Architect (Studio): International Studio (Crossboundaries, Beijing), Beijing Institute of Architectural Design Co., Ltd
Design: 2014-2015
Construction: 2014-2016
Floor Area: 26,000m²

地点：中国北京市
建筑师（事务所）：北京市建筑设计研究院有限公司国际工作室（Crossboundaries，Beijing）
设计时间：2014-2015
建造时间：2014-2016
建筑面积：26000m²

It is foreseeable that in the future schools in China will pay more attention to interaction, inspiration and individualization, presenting challenges to current educational ideas. Architects proposed a series of achievable redesign and adjustment to already-planned shell and core, thereby better supporting the school's long-term development.

To devise a stage for the passion of learning and teaching that will remain as the school and its pupils evolve, Architects reorganized classrooms and developed a program to include music and art subject areas. This was achieved by removing walls, adding windows and reconfiguring space. They all result from two key drivers: communication improvement and blurring of subject boundaries. To offer more lounge and sharing room, the whole space has been rearranged with a series of multipurpose elements, like functional walls, color identity, and both visual and physical connections.

What were institutional corridors and classrooms are now redefined spaces in terms of form and function. To encourage exchange and interaction between passing students, walls are integrated with lockers shelves and seating niches. Classrooms are reconceived to focus on students instead of teachers, to work across multiple subjects and activities. The classrooms' form breaks away from rigid uniformity, complemented by lighting, varying ceiling heights, a choice of vertical work surfaces and foldable walls to create different zones.

Passive and active connections engage the subjects of art, music and sport. The sport lobby bridges the double height space of art area, the ceiling imbedded with light strips mirrors the running track of the sports field above. Sports classrooms share circulation with the open plan art spaces and have windows with views directly into the sports halls. The library is now a crossroad, a successful partnering of the sculpting of the multifunctional hall ceiling with the above tribune, connecting the library and sport field.

可以预见的是，在未来中国的学校将更关注交互、启发与个性化的精神需求，而这将会对现行教育理念造成挑战。建筑师在学校已有的核心与表层设计之上，提出一系列可实现的调整和再设计，从而对学校的长期发展提供更好的支持。

在概念上，这更像是为教育和学习活动提供了一个平台，让学校和学生们在其间共同进步。为此建筑师对已有教室进行再组织并加入了例如音乐、艺术等主题的功能区域。具体手法则包括移除一些隔墙、加入开窗和对空间进行各种调整。一切设计都着眼于两个关键点：增强交流，以及模糊不同功能主题之间的界限。为了提供更多休息和分享的场所，整个空间被置入了一系列多用途元素，例如功能墙，颜色的一致性以及视觉与实体的双重连接。

习以为常的走廊和教室现在被重新定义了形式和功能。为了鼓励来往学生间的互相交流，墙体间被嵌入了储物柜和座位。教室则被重构为新的形式，使教学活动能更聚焦于学生而不是老师，并能兼容不同主题和活动。教室的形状不再是严格的统一，同时加入了变化的灯光与顶棚高度，以及不同的垂直工作平面与折叠墙，从而界定不同的区域。

主动与被动的联系渗入了诸如艺术、音乐和体育等不同的活动主题。体育大厅将艺术区通高的空间连接起来，同时顶棚上插入了与屋顶上跑道和运动场线条形状相对应的灯带。体育教室与开放式艺术空间共用动线，同时具有一组可以直接看向体育馆内部的开窗。图书馆则成为一个关键结点，同时也为通过屋顶连接图书馆与体育场的看台塑造大厅多功能顶棚提供了参考。

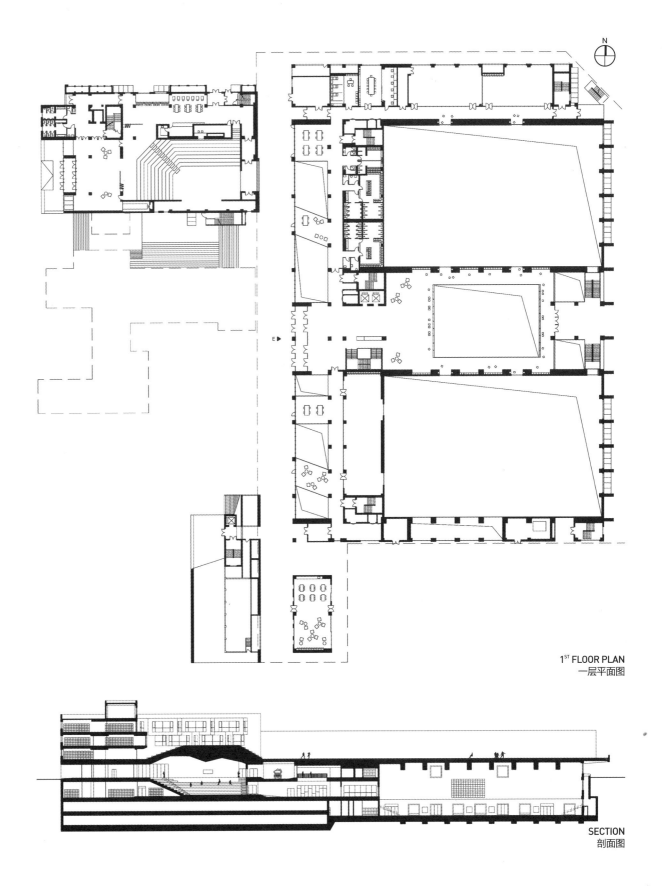

1ST FLOOR PLAN
一层平面图

SECTION
剖面图

Happiness Garden Exhibition Hall, Beichuan Earthquake Memorial Park

北川抗震纪念园 幸福园展览馆工程

Location: Beichuan County, Sichuan Province, China
Architect (Studio): Architectural Design and Research Institute of Tsinghua University Co., Ltd.
Design: 2009.06-2010.04
Construction: 2010.12
Site Area: 2,308m²
Floor Area: 2,353m²

地点：中国四川省北川羌族自治县
建筑师（事务所）：清华大学建筑设计研究院有限公司
设计时间：2009.06-2010.04
建造时间：2010.12
用地面积：2308m²
建筑面积：2353m²

The Beichuan Memorial Park project is located in the central part of the new town of Beichuan and is separated into three parts from west to east: Contemplation Garden, Hero Garden and Happiness Garden. As the main building of the Happiness Garden, the 13.6 meters high Exhibition Hall is located to the northwest of the Garden and covers an area of 2000m².

As for the single building of the Exhibition Hall, we strive to come up with a plan that allows it to contribute to a balanced layout of the overall design of the Memorial Park. Hence an asymmetric arrangement is adopted together with a withdraw from the park center and waterfront, making space of a square for people and activities. The roof is sloped properly enough for people to stay and move around. This paved incline is planted with trees and scattered with Z-shaped seats for resting. The main entrance of the Exhibition Hall is on the southeast side of the building. Main displaying space is on the -6.00m elevation while visitor service facilities and public service facilities are on the first floor (-2.00m elevation). These two floors are connected by gentle slopes. The two floors on the north side of the building are used as management offices. There is also a freight elevator next to the truck entrance.

The visitor circulation is set along the slope that centers along the city model in the main exhibition room. Visitors can circle around and overlook the model. The application of diverse tools like multimedia, photos, models and physical objects creates a modern and multi-level exhibition environment.

Local building materials are used. The facade features typical local material of blue stone and white stone that have well interpreted our design concept. Stones and woods are used in interior design as an extension of the external environment while creating a cordial atmosphere.

北川羌族自治县抗震纪念园选址于北川新县城中部，自东向西分为三部分，依次为静思园、英雄园和幸福园。展览馆位于幸福园西北侧，占地约2000m²，高度为13.6m，是幸福园的主体建筑。

展览馆建筑单体的布局力求在纪念园整体设计中求得均衡，因此在地段内采用非对称布局，面向园区中心及水面方向进行适当退让，留出广场人群活动的用地。建筑倾斜屋面坡度适宜，有利于人群停留活动。结合铺装，斜坡屋面种植了乔木，并设置了"之"字形休息座椅。

展览馆的主入口设在建筑东南侧，展陈空间主要集中在-6.00m标高层，上下两层主要以缓坡道连接。观众服务设施与公共服务设施设在首层（-2.00m标高）入口附近。管理办公等辅助用房设在北侧上下两层，货运电梯设在货运出入口附近。

展陈流线围绕主展厅城市沙盘延缓坡道排列，人流可环绕并俯瞰城市沙盘全景。同时结合多媒体、图片、模型、实物展等类型，创造出丰富、现代化的展览体验。

建筑单体主要使用当地材料，外部以青石、白石为主，以具有地方特点的材料搭配实现建筑设计的意图；室内使用石材和木材，将建筑外环境延续至室内，同时又不乏亲切之感。

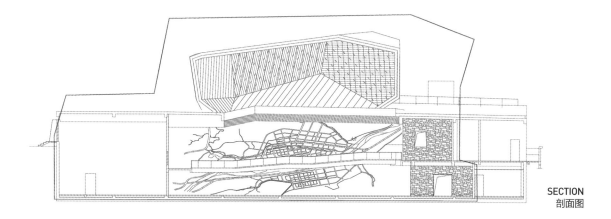

SECTION
剖面图

Creative Valley of South Taizi Lake (Phase I)

武汉南太子湖创新谷（一期）

Location: Wuhan, Hubei Province, China
Architect (Studio): CITIC General Institute of Architectural Design and Research Co., Ltd.
Design: 2012-2013
Construction: 2014-2017
Site Area: 31,816m²
Floor Area: 56,900m²

地点：中国湖北省武汉市
建筑师（事务所）：中信建筑设计研究总院有限公司
设计时间：2012-2013
建造时间：2014-2017
用地面积：31816m²
建筑面积：56900m²

The existing buildings in site were built in 1990's as light industrial plants. The project will become a symbol of revitalization of industrial heritage in a post-industrial context, by transforming the idle industrial structures into a vibrant multi-functional urban hub that integrates residential, commercial, cultural and recreational functions.

To enhance its architectural recognition while preserving as much as possible the historical structures, this design introduces dramatic architectural languages and innovative artistic functions which bring about not only the sentimental memories of the industrial past but also the dynamic reuse and regeneration of the heritage.

In our design, eight existing warehouses are grouped into two zones. The north zone will become museum of contemporary art, by integrating separated buildings with newly built structures. The museum has an open ground floor to face the city. Its two entrances are arranged in East and West Street corners. The large stairs will lead visitors up to the exhibition spaces in higher levels. The south zone is for retails and artist villages. It preserves the original layout and is designed into free spaces which could be allocated and assembled flexibly. Open public visitor routes are designed to activate different zones and enhance the collective memories of historical industrial spaces.

Creating a new mode of sustainable urban development, this project renews urban infrastructures and provides the city with more opportunities for major public activities by accommodating various art exhibitions and outdoor events. Promoting the communication between artists and the public, this project converts industrial heritage into one of the most dynamic and creative spaces in Wuhan.

汉阳，在中国清末的洋务运动中扮演着重要角色，是中国近代工业文明的发祥地之一。毗邻汉阳南面的沌口地区，是中国改革开放时期扩建的新兴工业区，拥有众多的中外合资兴办的现代企业。近年来，随着城市规模的扩张和人口的聚集，文体商贸及现代服务业等城市功能的复合度越来越高，城市定位逐渐开始由工业中心向复合型区域中心转变。

项目基地所在区域原是20世纪90年代建设的一片轻工业厂房，面对单调和缺乏特征性的建筑形态，我们试图通过植入新的艺术功能，以及充满张力的设计语言重新塑造富有戏剧性的建筑空间及路径体验，并最大限度地保留原有建筑的主体结构，实现既有建筑的更新与再利用。

基地原有的8栋厂房被分为两个大的区域。北侧的区域被设计为一座当代美术馆，通过加建部分区域将原来分散的几栋建筑连成一个整体。面向城市的部分，底层被全部打开，将人流引入场地内部。美术馆的入口则被安排在东西两侧街角，通过大台阶将游览路线引向高区的展览空间。南侧的区域作为商业及艺术家村落，基本保留原有建筑格局，被划分成可以灵活分配组合的自由空间，并通过开放的公共游览路径带动和激活各个区域，增强场所的记忆性。

建成之后的园区将作为公共性的文化建筑向市民开放，可以举办各类大型艺术展览或户外集会活动，创造更多市民与艺术家交流的机会，成为整个区域最具活力和创造力的场所空间。

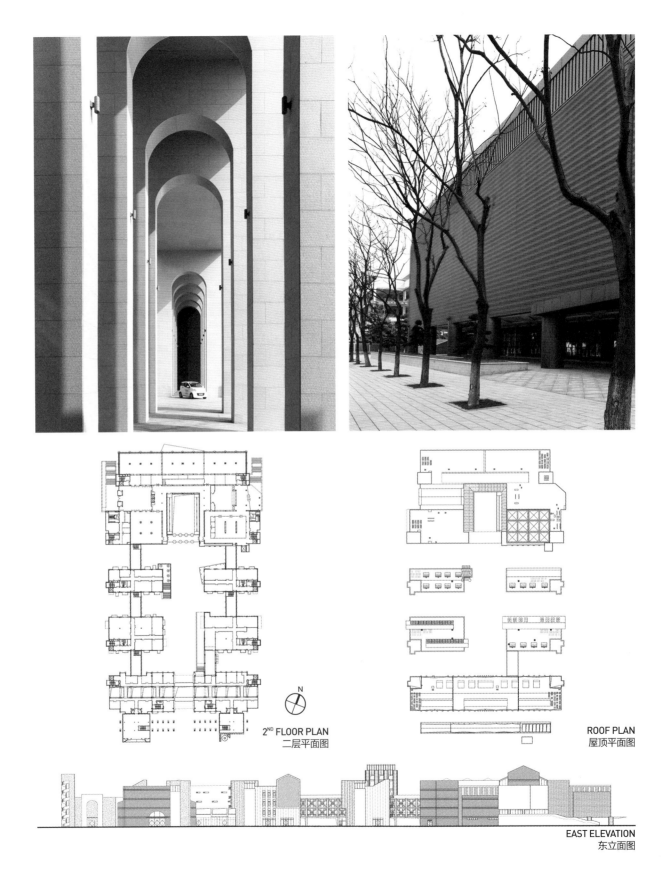

2ND FLOOR PLAN
二层平面图

ROOF PLAN
屋顶平面图

EAST ELEVATION
东立面图

China Welfare Institute Pujiang Kindergarten
中福会浦江幼儿园

Location: Pujiang New Town, Minhang District, Shanghai, China
Architect (Studio): Zhou Wei, Zhang Bin, Atelier Z+, Tongji Architectural Design(Group) Co.,Ltd.
Design: 2011.12-2014.03
Construction: 2013.06-2015.05
Site Area: 5,092m²
Floor Area: 15,329m²

地点：中国上海市闵行区浦江镇
建筑师（事务所）：同济大学建筑设计研究院（集团）有限公司致正建筑工作室：周蔚、张斌
设计时间：2011.12-2014.03
建造时间：2013.06-2015.05
用地面积：5092m²
建筑面积：15329m²

As a high standard supporting facility in Pujiang New Town, the China Welfare Institute Pujiang Kindergarten, consisting of twenty day-care classrooms and one early childhood education & teacher training center, it is located in the large area of low density communities.

China Welfare Institute has its own specific requirement on kindergarten space. Accordingly, the design of this project starts on the adequate response and guidance to the concerns of China Welfare Institute on space. As for the overall layout, the buildings are arranged on the north and east, leaving a large area of outdoor activity ground on the south and west. The building is presented as the two parallel micro-staggered laying teaching building in the northern-half of the base and a point-type preschool teacher training center in the southeast corner. They are connected as an entity by two bases with bottom floor accommodating all public activity facilities and management offices.

All the day-care classes are located at the south part of the second and third floor of these two teaching buildings. Apart from circulation and service facilities at the north, there is a corridor system with enlarged space which is equipped outside of each day-care classroom for extension of children's activities. There are also several small public spaces connecting different floors, making the dense classroom space of each floor extend quite a bit as well as enhancing the interaction between two floors.

As the kindergarten has an extraordinary configuration, how to control the sense of scale is a key for the project design. The selection of structure, material and color of building is another kind of extension for space and shape strategy.

作为浦江新镇的高标准教育配套项目，中福会浦江幼儿园由20个日托班和1个早教及师资培训中心组成，位于大片的低密度居住社区内。

中福会对幼儿园空间有明确的诉求，本项目就始于对这些诉求的充分回应与引导。总体布局上，建筑尽量靠北、靠东布置，留出南侧和西侧大片的户外活动场地。建筑整体呈现为基地北半部两栋平行微错布置的条形教学楼和东南角的一栋点式学前师资培训中心，它们由底层容纳了所有公共活动设施和管理办公的两个基座连成一个整体。

所有的日托班都在两栋教学楼的二、三层南侧，北侧除了交通、服务设施之外就是一个带有多处放大空间的走廊系统，每个日托班的活动室外都配有可以延展幼儿活动的放大走廊空间，并配有数个贯通上下楼层的小型共享空间，让每个楼层密集的班级空间在这些地方可以得到释放，同时也加强了楼层间的互动。

由于这个幼儿园的规模超过了一般配置，如何控制尺度感知成为设计中的一大重点。建筑的构造、材料与色彩选择其实也是空间与形态策略的延续。

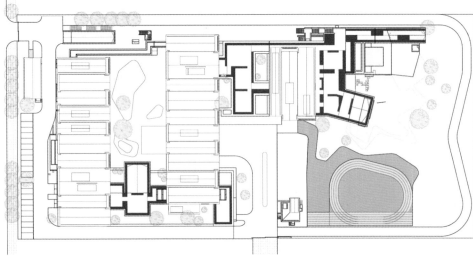

SITE PLAN
总平面图

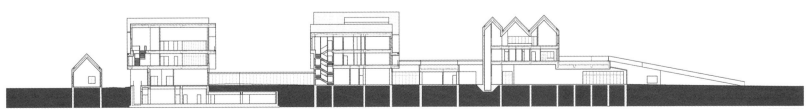

SECTION
剖面图

Humble Administrator's Villa

拙政别墅

Location: Suzhou, Jiangsu Province, China
Architect (Studio): East China Urban Architectural Design & Research Institute, Arcplus Group PLC
Design: 2010.05-2016.10
Construction: 2013-2016
Site Area: 42,326m²
Floor Area: 35,072m²

地点：中国江苏省苏州市
建筑师（事务所）：华建集团华东都市建筑设计研究总院
设计时间：2010.05-2016.10
建造时间：2013-2016
用地面积：42326m²
建筑面积：35072m²

Humble Administrator's Villa is located at one corner of the Humble Administrator's Garden. When our architects were designing it, we were expecting it to be a micro Suzhou rather than a common villa community.

RIVERS
Suzhou is called "the city of water" due to that rivers played very important roles in trafficthrough history. Our architects designed a landscape water system. All roads are sunken to create a river with no water. Such design is also a reflection of Suzhou water culture.

HILLS
These sunken roads give city life a feeling of living in the mountain. This landscape design creates a feeling that architectures are stood either by the flourishing tree, or on top of the mountain. All these above could offer people an unusual experience in Suzhou City.

BRIDGES
The pedestrian circulation connects the entrance to every villa and the community's landscape meticulously. It has twelve arch bridges crossing the river without water and each of them represents a different month of a year. That's why we call them Twelve-Month-Bridges.

GARDENS
The whole community is divided into 29 buildings, and each of them has its own individual garden. The whole master landscape design of this community could be seen as a gigantic classic Suzhou garden. Humble Administrator's Villa is indeed a reincarnated newly born magnificent Suzhou garden inside a unique group of gardens.

苏州拙政别墅地处拙政园一隅，设计师希望这是个微型的苏州，而不只是一个别墅区。

河流：
苏州为水城，历史上的河流承担了交通功能，设计师不仅沿着步行空间构筑了贯穿全园的景观水道作为对苏州水文化的映射，更将道路下沉，将承载现代交通工具的下沉道路视作旱河，也体现一种对苏州水文化的追忆。

山林：
下沉的道路不仅将消防交通以及地库的入口综合在一起，更无意中创造了城市山居的效果。沿着垂直高度结合步行系统被设计成精致的坡台景观，让建筑仿佛悄然设立在花树之侧或之上。从而营建出苏州城内别样的山林意趣。

桥梁：
小区的步行系统将各幢别墅的入口以及小区景观精细地联系起来，有十二座拱桥跨越旱河，这十二座拱桥对应一年十二月，故名十二月桥。

园囿：
小区的各个别墅拥有独立的园林，别墅的园子和公共的景观在设计上有相互暗示、借景以及联系，而园区的公共景观则是个大园子。故而拙政别墅是个园中复园的超级苏州园林。

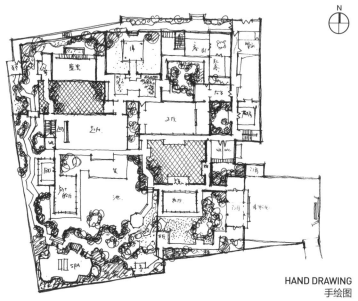

HAND DRAWING
手绘图

HAND DRAWING
手绘图

Customs Clearance Service Center of Chenglingji Free Trade Zone in Hunan

湖南城陵矶
综合保税区
通关服务中心

Location: Yueyang, Hunan Province, China
Architect (Studio): Su Chang, Tan Chunhui, Institute Of Shanghai Architectural Design & Research (Co., Ltd.), Arcplus Group PLC
Design: 2015
Construction: 2016
Site Area: 36,935m²
Floor Area: 49,250m²

地点：中国湖南省岳阳市
建筑师（事务所）：华建集团上海建筑设计研究院：苏昶、谭春晖
设计时间：2015
建造时间：2016
用地面积：36935m²
建筑面积：49250m²

Functional areas, such as office, public service, customs declaration hall, conference and exhibition center and finance office, are integrated into a building with 5 separate parts. Each part is separated from the others, creating some heterogeneous voids defined as sharing spaces. The three-storey central hall joins these voids together into a connecting system, where air can flow everywhere and the whole space is flowing, too. The isolation between the building and the environment as well as between people is eliminated, creating a humanized public service building with a more open and more participatory atmosphere around.

The combination of slope-shaped green roofs and platforms will create a multi-level and rich activity space, which echos and extends the internal sharing space. The vision far away is showing through the terrace on the platform of the 4th floor, where a stunning landscape of Bajiao Lake is coming into people's eyes directly.

The façades are equipped with the most effective shading systems aiming at different orientations, which become a part of elegant façade through delicate design. Utilization of passive technology and recycling of waste heating steam forms a large CCHP energy-saving mode.

办公、便民服务、报关、会展、金融等使用需求被整合为5个独立的功能实体，将这五个实体相互分离产生不均质的虚空，这些虚空被定义为共享空间。三层通高的中央大厅将这些空间串联起来——空气是流动的，空间也是流动的，建筑与环境之间、人与人之间的隔绝被消解了，创造出具有参与性、开放性的人性化公共服务性建筑。

近地楼层坡形屋面绿化与平台结合，营造出多标高、层次丰富的活动空间，与内部共享空间形成良好的延续和沟通。借景的精心组织，展示了令人震撼的芭蕉湖山水长卷。

各立面针对不同朝向分别采用最有效的遮阳系统，通过巧妙设计，使其成为精美立面的一部分。大量被动技术的采用，并利用火电厂废热蒸汽形成大"三联供"的节能模式。

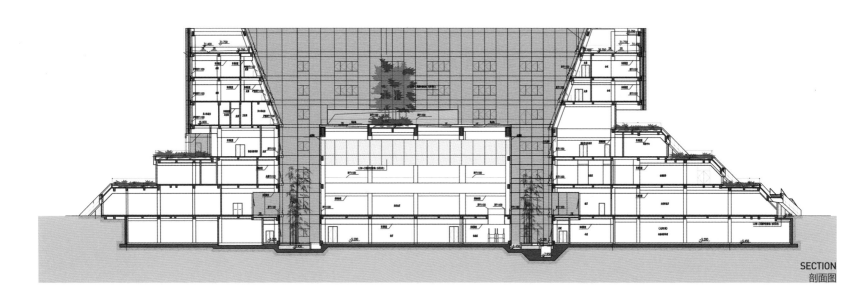

SECTION
剖面图

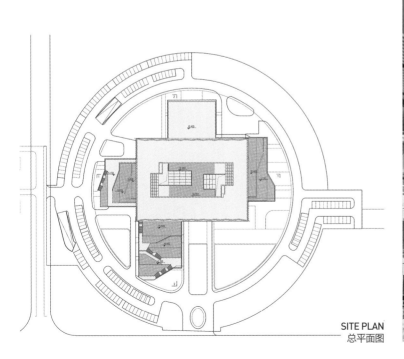
SITE PLAN
总平面图

WEI Retreat Tianmu Lake

天目湖微酒店

Location: Liyang City, Jiangsu Province, China
Architect (Studio): Institute Of Shanghai Architectural Design & Research (Co., Ltd.), Arcplus Group PLC
Design: 2013.09-2014.07
Construction: 2014.07-2016.04
Site Area: 7,385m²
Floor Area: 8,998m²

地点：中国江苏省溧阳市
建筑师（事务所）：华建集团上海建筑设计研究院
设计时间：2013.09-2014.07
建造时间：2014.07-2016.04
用地面积：7385m²
建筑面积：8998m²

WEI Retreat Tianmu Lake is located at 8 Huanhuxi Road, Tianmu Lake tourism resort, Liyang City, Jiangsu. It is surrounded by the Tianmu Lake in the east, west and south, and West Huanhu Road in the north. We have designed functional areas including F&B, accommodation, and supporting logistics according to land use planning conditions, as an exclusive project for enterprise customers and high-end tourists, so as to create a building group with reasonable layout, comprehensive facilities, pleasant spatial dimension, and distinct architectural individuality. While catering to the cultural and consumption needs of this tourism destination, it has become a unique boutique hotel with rich connotation. This understated hotel has only 35 guest rooms across 2 floors, covering an area of only 4800 square meters. Yet it is a luxury hotel fully embracing the whole view of South Mountain, overlooking Tianmu Lake, and hosting a natural hot spring. The natural beauty of lakes and mountains as well as the spirit of heaven and earth seems to be concentrated in the small hotel, just like the boundless universe contained in the dust.

"Oriental flavor by the water side." This project enjoys a prime location to the north shore of Tianmu Lake on an open peninsula, with a panoramic view of the lake. Taking into account of the concept of fully protecting water resources, the hotel was integrated into the whole ecology and natural scenery. The building uses courtyards to connect a series of spaces themed with oriental culture, and uses concise lines, pleasant scales and exquisite details to create the special beauty of elegance and refinement in oriental culture.

The special courtyard-like entrance provides enhanced visibility from the outside, and its simple yet well-considered landscape layout has reflected the unique style of oriental courtyard layout. The combined design of meticulous lighting and wall moldings and even the Chinese style pane outside of window, has formed the transitional spaces before entering the main lobby. As you walk inside, you will discover lush gardens, and free-flowing music-like architectures.

The iconic infinity pool, combined with the U-shape layout of buildings, has formed the theme of inner courtyard. The green space and vegetation of all seasons have further created a unique spatial experience, making guests feel like walking in natural sceneries of lakes and mountains.

天目湖微酒店位于江苏溧阳市天目湖旅游度假区环湖西路8号，东南西三个方向邻天目湖区，北临环湖西路。根据用地规划条件和使用因素，设置餐饮、住宿及配套后勤等功能板块，作为一个面向企业客户群及高端旅游人群的会馆项目，营造一个布局合理、设施配置齐全、空间尺度宜人、建筑个性鲜明的建筑群落。在补充完善本地区的文化消费等功能的同时，成为区域内独一无二、内容丰富的精品酒店。在这一片净土上的这座小小的酒店，小到只有35间客房，只有2层楼，地上只有4800m²。但它却又是一座大酒店，大到它拥有了整座南山，望见了整个天目湖，坐享了整片的天然温泉。这一片湖光山色，天地灵气仿佛都浓缩在这座小小的酒店里。犹如微尘之中蕴含的大千世界。

"东方韵味，在水一方"——本项目地处天目湖北岸的中心位置，视野开阔，呈半岛式布局，独有欣赏湖区全景的优势。设计在充分保护水资源的前提下，融入整个生态与自然景色中去。建筑用庭院串联一系列以东方文化为主题的空间，用简洁的线条、宜人的尺度、精致的细部，演绎出东方文化特有的优雅精致之美。

特色的入口庭院——最大限度地延伸了室外入口的视线范围，景观的布置，简洁精致，体现了东方院落布局的建筑个性。细致的灯光与墙体线脚的融合设计乃至窗外的中式窗格都如江南风景的秀色点缀，形成了进入主体大堂前的过渡层次。越往里走，满园春色渐出，建筑也如音乐般随性华美。

特色的无边水池——结合U字形建筑围合的布局，形成了建筑内院的主题。无边水池将宾客的视线引向远方，与湖光天色相接，把内院的景观无限地延伸与扩大了。绿地和四季植被景观，进一步营造了独特的空间体验，使行走其中如同置身湖光山色之中。

HAND DRAWING
手绘图

Shooting Range Hall of the East Asian Games

东亚运动会射击馆

Location: Tianjin, China
Architect (Studio): Tianjin Architecture Design Institute
Design: 2010-2011
Construction: 2011-2013
Site Area: 74,811m²
Floor Area: 37,996m²

地点：中国天津市
建筑师（事务所）：天津市建筑设计院
设计时间：2010-2011
建造时间：2011-2013
用地面积：74811m²
建筑面积：37996m²

The Shooting Range Hall is one of the specified venues for the East Asian Games. It is comprised of four separated venues, and totally 2665 seats, including 10-meter, 25-meter, 50-meter-range preliminaries and one for final competition.

Designed in accordance with the Rules and Regulations of ISSF, it meets national standards put forward by General Administration of Sport of China, and can undertake for hosting both national and individual international event.

Its architectural layout fits the flow of audience, athletes, coaches, event managers, referees and media. It has the features like clear division, convenient lines of flow and well-equipped functions.

Its architectural modeling design takes "speed" as its theme to evoke a vigorous and forcible unique dimension by staggering motifs with different scales and volumes.

Parametrically modeled building facades improve the completion and construction operation of nonlinear curtain walls.

Post-game utilization is considered in the design to integrate theaters, table tennis courts, badminton courts and other functions, and improve the efficiency of utilization of sports buildings.

射击馆是东亚运动会的专项比赛场馆之一，功能包括10m、25m、50m预赛馆和决赛馆，总观众席位2665座。场馆设计符合《国际射联章程与规则》标准，达到国家体育总局射击射箭运动管理中心的使用要求，可承办全国性和单项国际比赛。

建筑布局合理组织观众、运动员、教练、赛事管理、裁判、媒体等多种流线，分区明确、流线简捷、功能完备。

建筑造型设计以"速度"为主题，用不同尺度和体量的母题交错形成刚劲有力的独特体量。建筑立面采用参数化设计，提高了非线性幕墙设计的完成度和施工操作性。

设计综合考虑场馆赛后利用，融合剧场、乒乓球场、羽毛球场等功能，提高体育建筑使用效率。

天津体育中心射击馆
TIANJIN SPORTS CENTER SHOOTING RANGE

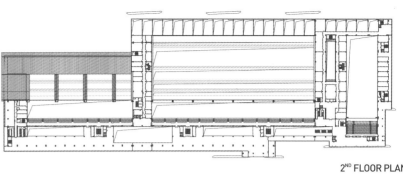

2ND FLOOR PLAN
二层平面图

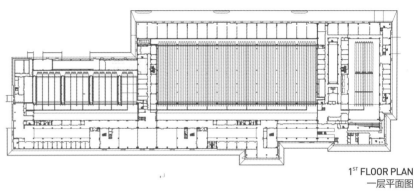

1ST FLOOR PLAN
一层平面图

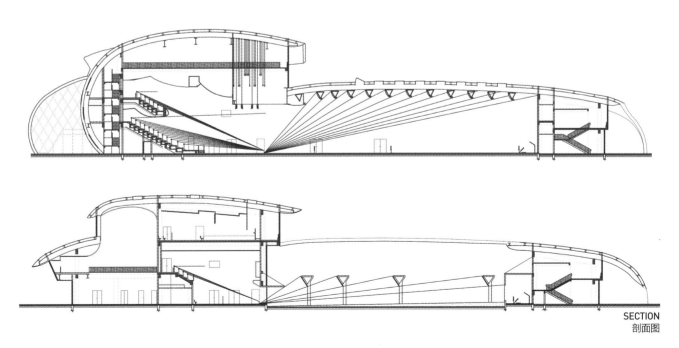

SECTION
剖面图

Yichang Planning Exhibition Hall

宜昌规划展览馆

Location: Yichang, Hubei Province, China
Architect (Studio): East China Urban Architectural Design & Research Institute, Arcplus Group PLC
Design: 2013.05-2014.12
Construction: 2016.02
Site Area: 29,983m²
Floor Area: 20,960m²

地点：中国湖北省宜昌市
建筑师（事务所）：华建集团华东都市建筑设计研究总院
设计时间：2013.05-2014.12
建造时间：2016.02
用地面积：29983m²
建筑面积：20960m²

The Planning Exhibition Hall as an important landmark of Yichang is located in the heart of the new district, and the architecture echoing surroundings, is formed like a rolling mountain. The Landscape design continues the concept, echoing with the building with large pitched grass ramps. The building focuses on strengthening the connections among people, architecture and the environment, and make visitors walking in and outside the building feel like climbing a mountain with extremely fun. The overall image of the building is presented from three different views as below:

Distant view: the contour of the building echoes with the rolling, boundless mountain, and also it is a new interpretation of traditional Chinese pitched roof.

Middle view: the wedge-shaped mass is in sharp contrast to the glass box of vertical louvers: Moreover, the design of inner courtyard and roof garden broadens the architecture space and enriches the levels of landscape.

Close view: the material of outer skin is applied with double layer perforated aluminium alloy panels, and it sinner layer is in gold colour, which reflects the aesthetics of modern mechanical and exquisite details.

宜昌规划展览馆作为宜昌市重要的地标建筑，位于宜昌市新区的核心区域，建筑形式与周边环境相呼应，仿佛层峦叠嶂的山体。景观设计延续山体形象这一概念，用大面积斜面草坡与建筑呼应。建筑强调人、建筑与环境之间的联系，使得游人行走于建筑内外，犹如爬山观景，别有一番趣味。建筑的整体形象从以下三个不同视野进行阐述：

远景：建筑的外轮廓与连绵起伏的山体相呼应，同时又是中国传统元素坡屋顶的新的解读；

中景：楔形体量和装有竖向百叶的玻璃盒子形成鲜明的虚实对比，内庭院和屋顶花园的设计又丰富了建筑的空间和景观层次；

近景：建筑外表皮采用双层金属穿孔铝板，内层金色，外层渐变穿孔铝板，体现出颇具现代感的机械美学和精致的细部。

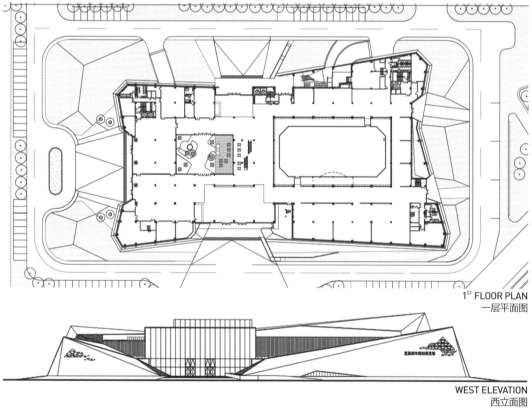

1ST FLOOR PLAN
一层平面图

WEST ELEVATION
西立面图

Liu Haisu Art Museum

刘海粟美术馆

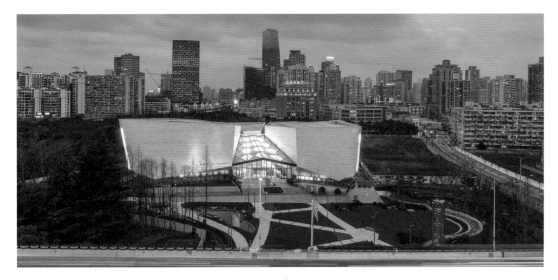

Location: Shanghai, China
Architect (Studio): Tongji Architectural Design(Group) Co.,Ltd.
Design: 2012
Construction: 2015
Site Area: 6,000m²
Floor Area: 12,540m²

地点：中国上海市
建筑师（事务所）：同济大学建筑设计研究院（集团）有限公司
设计时间：2012
建造时间：2015
用地面积：6000m²
建筑面积：12540m²

Located in Shanghai downtown, Liu Haisu Art Museum is a national art museum named after Mr. Liu, the founder of China's modern art education.

Notwithstanding limitations of available land, the building is expanded at the top to allocate as much landscape space from the ground floor as possible and maximize social benefits.

The design of the art museum is inspired by Mr. Liu Haisu's profound life and art accumulation. The design idea, "sea of clouds and mountain of rocks", is from the imitation of sheer rocks of the Mount Huangshan that is both the friend and the teacher for Mr. Liu in his lifetime. This idea is also a permanent theme of Chinese picturesque landscape painting. The rough cutting of big volume on the cube brings a strong sense of sculpture to represent Mr. Liu's crazy and relentless pursuit in arts. The smooth integration of modern and traditional, eastern art and western aesthetics is the lifelong pursuit of Mr. Liu and the aim of the design.

Through the rough cutting of big volume on the cube brings a strong sense of sculpture to represent Mr Liu's crazy and relentless pursuit in arts. Concise and powerful folding interfaces draw the outline of building volume, but also endow the building with dynamics. The division of skylight follows the Chinese traditional style of rafters and eaves, and integrates Mr. Liu Haisu's lifelong traditional Chinese and Western artistic achievements through traditional artistic conception and modern architectural language.

坐落于上海市中心的刘海粟美术馆，是以中国现代美术教育的奠基人之一刘海粟之名命名的国家重点美术馆。

用地局促的条件下，美术馆底层仍尽可能多分配为绿地以实现社会效益最大化，将建筑往空间扩展。

美术馆的设计从刘海粟先生深厚的人生与艺术积淀中汲取灵感，同时以折线的建筑语言呼应其亲自选择的原美术馆方案气质。设计立意"云海山石"取意于刘海粟一生"为师为友"的黄山，也是国画永恒的主题。而大开大阖的形体则塑造出强烈的现代艺术雕塑感。现代与传统、东方艺术与西方美学有机地融合，既是刘海粟一生的艺术成就，更是刘海粟美术馆的设计理念。

设计通过大手笔的体量切割表现强烈的雕塑感，把刘海粟先生对艺术的不懈追求固化为充满活力的美术馆。简洁有力的折面勾勒出大气的建筑体量，亦为建筑赋予强烈的动感。中庭玻璃天窗的分隔与走向沿袭中国古典建筑的椽与屋檐，将刘海粟先生一生横贯中西的艺术成就通过传统意境的塑造以及现代建筑语言的雕琢有机融合。

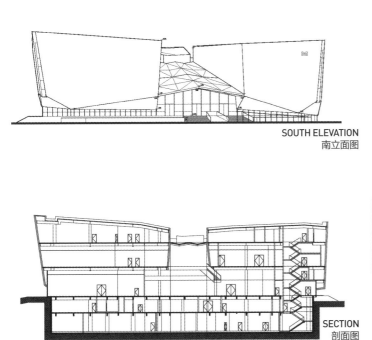

SOUTH ELEVATION
南立面图

SECTION
剖面图

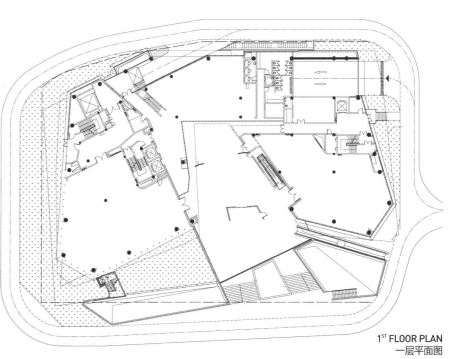

1ST FLOOR PLAN
一层平面图

Shenzhen Maritime Sports Base and Marine Navigation Sports School

深圳海上运动基地暨航海运动学校

Location: Longgang District, Shenzhen, Guangdong Province, China
Architect (Studio): Beijing Institute of Architectural Design Co., Ltd
Design: 2009.04-2010.04
Construction: 2011.04
Site Area: 81,800m²
Floor Area: 27,180m²

地点：中国广东省深圳市龙岗区
建筑师（事务所）：北京市建筑设计研究院有限公司
设计时间：2009.04-2010.04
建造时间：2011.04
用地面积：81800m²
建筑面积：27180m²

The project is located in Judiaosha Area, Longgang District, Shenzhen, China, and is composed of plot A (Shenzhen Maritime Sports Base) and plot E (Marine Navigation Sports School). The base area is divided into seven functional areas according to the technological process and the requirements of the 26th World University Games. They are operation area, sports competition area, athletes and team officials area, concierge area, TV broadcasting area, media area and audience activity area. The venue is for sailing tournament of the Games, and it will be changed into Sports Center of Shenzhen after the Games. In plot E there are education building, students dorms, teachers' apartments, canteen and indoor sports stadium. During the Games, it is the athlete village for maritime sports events and will be the Marine Navigation Sports School of Shenzhen after the Games. Although the total area is 27,180m², it has multiple functions.

According to local conditions, the architecture fully fit to the surrounding terrain and the environment. The buildings are arranged along the harbour are arranged in arc line, thus naturally showing the well-proportioned architectural image which is distinct both vertically and horizontally, and ensuring a good perspective. It is formed into an organic ecological group of constructions in the area.

本项目位于深圳市龙岗区南澳街道新东路桔钓沙片区，由A地块（深圳海上运动基地）及E地块（航海运动学校）两部分组成。其中A地块基地场区按帆板比赛工艺流程及大运会赛会组织需求共划分为七大功能区域，分别为场馆运营区、体育竞赛区、运动员及随队官员区、场馆礼宾区、电视转播区、新闻媒体区及观众活动区。赛时为第26届世界大学生运动会帆板比赛的竞赛设施，赛后为深圳市海上运动中心。E地块包含有教学楼、学生宿舍楼、教师公寓、食堂及室内运动馆。赛时为大运会海上运动项目专用的运动员分村，赛后为深圳市航海运动学校。整个项目虽然总建筑面积仅有27180m²，但其各类功能区域众多，各场地内的建筑布局均依山就势，沿现状港湾呈弧形排列，保证了所有建筑均能有良好的视野景观，整个建筑组群自然地融入场地环境，在深圳市大鹏湾地区形成了一处富于韵律的有机生态组群。

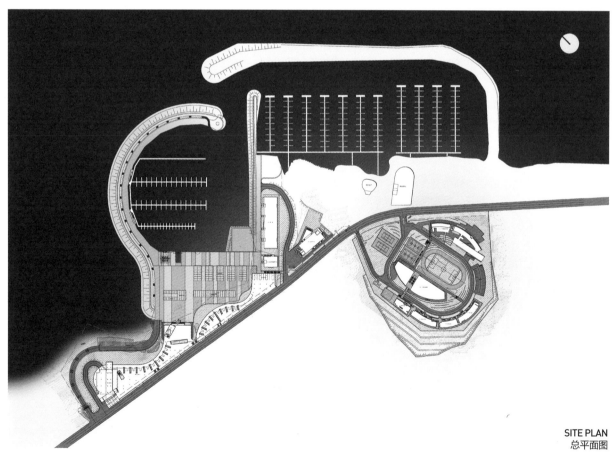

SITE PLAN
总平面图

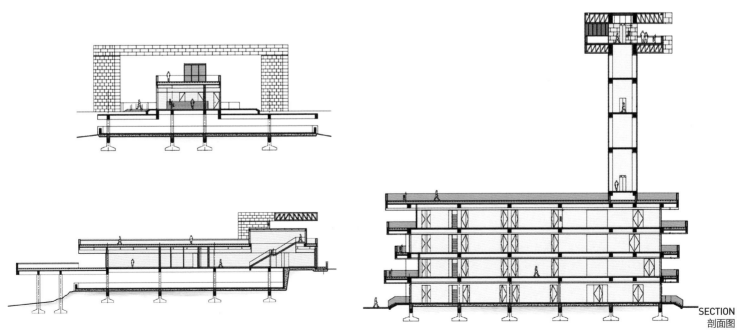

SECTION
剖面图

Shandong Art Gallery 山东美术馆

Location: Lixia District, Ji'nan, Shandong Province, China
Architect (Studio): Li Li, Tongji Architectural Design (Group) Co., Ltd.
Design: 2011.9-2012.9
Construction: 2012.1-2013.8
Site Area: 20,700m²
Floor Area: 52,138m²

地点：中国山东省济南市历下区
建筑师（事务所）：同济大学建筑设计研究院（集团）有限公司；李立
设计时间：2011.9-2012.9
建造时间：2012.1-2013.8
用地面积：20700m²
建筑面积：52138m²

The gross floor area of the new gallery is 52,000m², including five floors above ground and one floor underground. 12 exhibition halls are arranged in the art gallery from the first floor to the fourth floor. The total area of these halls is 19,700m² and the available exhibition line is 1,600 meters long.

In order to meet the requirement as an important spatial node for eastbound development of Jinan, the Shandong New Art Gallery located in Shandong Culture Expo Center is required to become the leader and integrator for this space. On one hand, the art gallery needs to echo with the square shape of museum, archives room and the axis relationship, in order to form "品" shape structure. On the other hand, the art gallery needs to create square in the front together with the museum in order to form an interface of some length for the square. The body characterizes with 4 floors in its main part and partially 6 floors in some other parts is established in the design.

In order to echo the surroundings and geographical features of Jinan city, the theme that the mountain and the city lean with each other is established in the architectural design of art gallery. The building body transits from 4-floor mountain shape to 6-floor square shape. Under the concept of "mountain and city lean with each other, spring and city reflects with each other", the style of the building utilizes backward step shape from south to north. The number of floors gradually increases toward north. The shape changes gradually from trapezoid to cube while the skylight reflects the image of "spring".

The image of Chinese traditional garden is also integrated in the design. As the scale and space of the art gallery are limited, the trapezoid turning space is designed on the south side. It corresponds to two gaps on the exterior of the building and enhances diagonal space. In addition, it allows every entity that participates space construction to be dislocated and twisted relative to each other, thereby enhancing the sense of spatial flow.

新美术馆的建设总面积5.2万m²，其中地上五层，地下一层。馆内一至四层设有展厅12个，展览总面积19700m²，实际可用展线1600m。

为了符合济南城市向东发展的重要空间节点，位于山东文博中心的山东省美术馆新馆需要成为这一空间的引领者和整合者。一方面，美术馆要与博物馆、档案馆的方整体量以及轴线关系取得呼应，形成"品"字形格局；另一方面又要与博物馆共同围合馆前广场，形成一定长度的广场界面。于是，设计确立了美术馆主体四层、局部六层的体型特征。

为了呼应周围环境及济南地理特点，美术馆的建筑设计确立了以山、城相依的主题，建筑形体从四层的山形自然过渡到六层的方形；并在"山·城相依、泉·城相映"理念的贯穿下，场馆样式从南往北采取了退台式形体，往北层数渐渐增高，形状由梯形渐变为正方体；而"泉"的意象则由屋顶的天窗来体现。

设计中还融入中国传统园林的意象。美术馆的范围、空间也有限制，因此设计了在南面的梯形空间转折的位置，对应建筑外观的两个缺口，并在此强化对角空间；让每一参与空间构建的实体彼此间错动、扭转，加强空间的流动感。

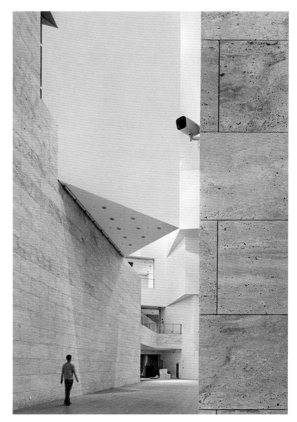

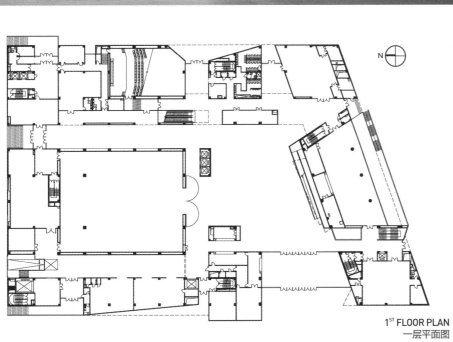

1ST FLOOR PLAN
一层平面图

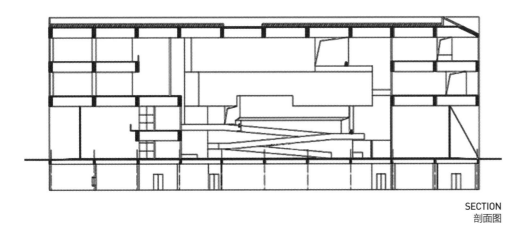

SECTION
剖面图

D23 Project, Plot 8 of Hongqiao Business District

虹桥商务区
8 号地块
D23 项目

Location: Shanghai, China
Architect (Studio): East China Architectural Design & Research Institute, Arcplus Group PLC
Design: 2012.12
Construction: 2014.07
Site Area: 43,710m²
Floor Area: 230,000m²

地点：中国上海市
建筑师（事务所）：华建集团华东建筑设计研究总院
设计时间：2012.12
建造时间：2014.07
用地面积：43710m²
建筑面积：230000m²

The project is located in D23 neighbourhood of plot 08, the core area of Hongqiao business district (phase I). The plot consists of 7 parcels.

1. Rational Urban Scale

Inspired by similar elements in spatial enclosure and urban scale of Berlin Potsdam's classic small-scaled & mixed-use blocks and Shanghai's older downtown area, the design creates a brand new business community with continuous urban interface and multi-layered public indoor spaces for socializing.

2. Shared Courtyards and Independent Spaces

While the landscape and partial service facilities are open to the public, the design has fully considered the independence demand of each headquarters.

3. Double Halls — Separated Circulation for Walkers and Cars

The sunken courtyards introduce daylight and vegetation to basement office entrance halls, echoing ground-floor entrance halls.

4. Vertical Greening

The sunken courtyards introduce greening to commercial service facilities and office entrances in B1 and B2, and turn them into delightful spaces. The multi-layered vegetation forms a three-dimensional landscape system.

5. Flexible Layout Arrangement

Large span spaces are applied in office building. The column-free large spaces are flexible either as a whole or re-divided for office, meeting or other uses, greatly increasing efficiency.

6. Building Form and Details

The material use in elevation highlights contrast of solid and void. The light and vivid building form incorporates delightful and transparent offices.

D23项目坐落于虹桥商务区核心区（一期）08地块D23街坊，整个地块划分为7个小地块。

1. 宜人的城市尺度

借鉴"经典的柏林波兹坦小尺度混合街区与上海老城区相类似的空间围合和城市尺度关系"，打造有着完整城市界面以及丰富内部公共交往空间的商务社区。

2. 公共庭院、独立空间

尺度宜人的建筑体量充分考虑地块景观及服务设施公众性，又兼顾总部办公独门独户需求。

3. 双大堂理念——人车分流

下沉庭院的办公入口门厅有良好的采光和景观，与地面入口门厅共同形成丰富的办公楼入口空间。

4. 立体绿化环境

设置下沉庭院，为地下一、二层商业服务设施及办公入口提供犹如地面的绿化环境，通过多层次绿色植被形成立体的绿色景观体系。

5. 灵活的建筑平面布局

以大跨度大空间作为办公楼的基本构架，既可实现无柱大空间灵活高效的办公、会议模式，也可便利地作各种功能空间分隔。

6. 建筑形态与细部处理

建筑外立面虚实结合，形成通透明快的办公环境和轻盈灵动的建筑形态。

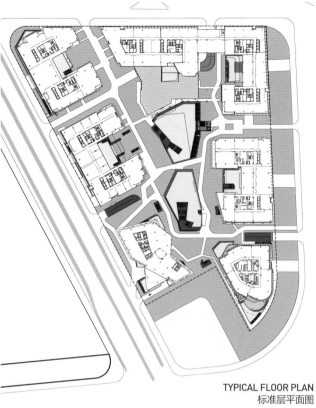

TYPICAL FLOOR PLAN
标准层平面图

PERSPECTIVE DRAWING
透视图

Aimer Fashion Factory 爱慕时尚工厂

Location: Shunyi District, Beijing, China
Architect (Studio): International Studio(Crossboundaries, Beijing), Beijing Institute of Architectural Design Co., Ltd
Design: 2004
Construction: 2013
Site Area: 62,236m²
Floor Area: 53,000m²

地点：中国北京市顺义区
建筑师（事务所）：北京市建筑设计研究院有限公司国际工作室（Crossboundaries，Beijing）
设计时间：2004
建造时间：2013
用地面积：62236m²
建筑面积：53000m²

Its single large volume representing the company image cultivates and communicates the message of brand as an industry leader from architecture to interior design and to site design. Aimer Fashion Factory is a place for people to work rather than simply focusing on factory elements.

Established in 1992, Aimer Lingerie is the industry leader producing high-end lingerie, underwear and body wear. The complex houses their production lines, distribution center, storage facilities, quality control, R&D departments, as well as department offices.

One continuous story is told across the whole complex, from the first impression given by the size of building when we are approaching, to the landscape that surrounds it and tears through the centre of the mass as a cavernous void that brings light and landscape into the depth of industrial building.

In order to reinforce the impression of brand, natural landscape isn't just restricted to the outside. Instead, it continues as a journey inside. Translated from the architectural language of a larger scale, a landscape of paths and patterns creates communication among departments and maintain transparency to visitors.

代表企业形象的大型单独体量，从建筑设计到室内设计再到场地设计，无不培养和传递着作为行业领导的品牌信息，爱慕时尚工厂的设计与其说是着眼于一座工厂，不如说是营造一个供人们工作的场所。
成立于1992年的爱慕内衣集团是行业内的领导者，一直从事高端女士内衣、内裤及贴身衣物的制造。这座综合体建筑容纳了他们的生产线、物流分发中心、仓库、品控、研究和开发部门，同时还有不同部门的办公室。
整个综合体旨在创造一个统一的感受，具体表现从靠近建筑时的整体体量给人造成的第一印象，到其周围的景观设计，一直到穿过体量正中被分开、将自然光引入工业建筑深处的峡谷状的空间，连续的楼层平面贯穿整个综合体。
为了加深这种印象，自然景观也没有被局限在建筑外部，而是被引进了室内。通过对更大尺度上的建筑语言进行再演绎，由小路与图案组成的室内景观营造了不同部门之间的交通联系，同时为参观的访客保留了充分的透明度。

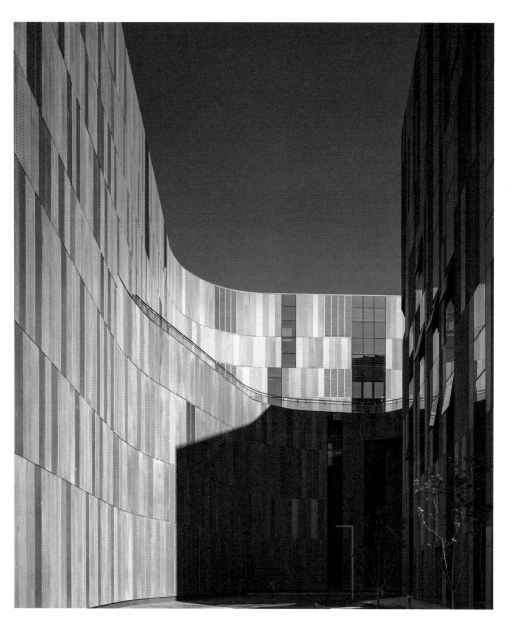

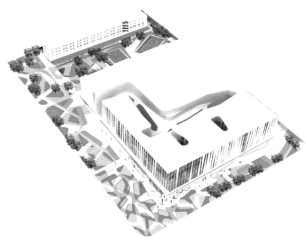

PERSPECTIVE DRAWING
透视图

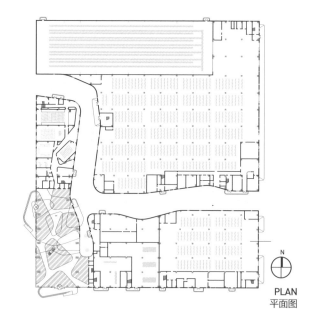

PLAN
平面图

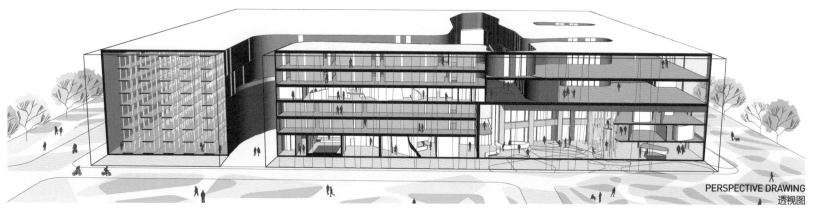

PERSPECTIVE DRAWING
透视图

2022 The Winter Olympics Plaza

2022 首钢西十冬奥广场

Location: Shougang Garden, Shijingshan District, Beijing, China
Architect (Studio): CCTN Architectural Design Co. Ltd.
Design: 2016.03
Construction: 2016.04
Site Area: 76,952m²

地点：中国北京市石景山首钢厂区
建筑师（事务所）：中联筑境建筑设计有限公司
设计时间：2016.03
建造时间：2016.04
用地面积：76952m²

By splitting space and adding units based on the original industrial layout, we redesign to transform enormous structures scattered around into a pleasant, vibrant pentagonal courtyard that is livable and connected for its inhabitants, creating the oriental philosophy of architecture that emphasizes on the beauty and advantage of "community".

The 2022 Winter Olympics Plaza is located at the northwest corner of the site of the old Shougang plant, with floor area of 87,000m², designed as offices, conference centers, exhibition spaces and peripheral facilities. The old structure is scrupulously preserved to retain its original architectural features along with the newly-added units, paying respect to its industrial past. Outdoor stairs and corridors crossing buildings and rooftops are built to retain the authentic industrial design while enriching with the features of classical Chinese garden. The whole group of buildings is a stereoscopic industrial landscape where unique Chinese spatial dynamics are conveyed as the scenery changes when your steps are moving on.

设计通过众多近人尺度的插建和加建细腻地缝合了原有基地内散落的工业构筑物，营造出了景色宜人、充满活力的院落，以"院"的形式语言回归了东方最本真的关于"聚"生活态度。

2022首钢西十冬奥广场位于首钢旧厂址西北角，总建筑规模约8.7万m²，主要为办公、会议、展示及配套。设计谨慎保留原有建筑结构，造型忠实呈现出了"保留"和"加建"的不同状态，表达了对既有工业建筑的尊重。穿行于建筑之间和屋面的室外楼梯及步廊系统为整个建筑群在保持工业遗存原真性的同时，叠加了园林化特质。整组建筑就是一个立体的工业园林，步移景异间，传递出一种中国特有的空间动态阅读方式。

SITE PLAN
总平面图

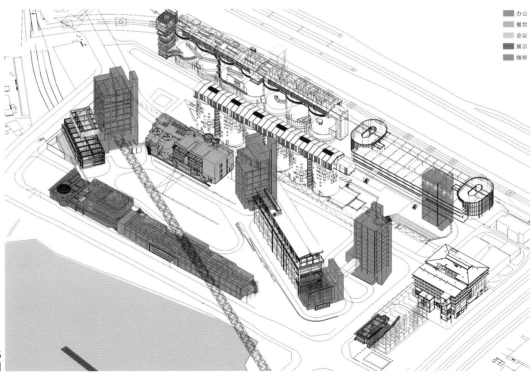

FUNCTIONAL ANALYSIS
功能分析图

Xi'an North Station of Zhengzhou-Xi'an High-speed Railway

郑西高铁西安北站

Location: Weiyang District, Xi'an, Shaanxi Province, China
Architect (Studio): Tang Wensheng, Central-South Architectural Design Institute Co., Ltd.
Design: 2008.09-2009.07
Construction: 2008-2010
Site Area: 439,043m^2
Floor Area: 332,000m^2

地点：中国陕西省西安市未央区
建筑师（事务所）：中南建筑设计院股份有限公司；唐文胜
设计时间：2008.09-2009.07
建造时间：2008-2010
用地面积：439043m^2
建筑面积：332000m^2

Located on the north end of the central axis of Xi'an, Xi'an North Station is one of the six major railway passenger hubs. The passenger line of "entering from top and leaving from bottom, stereoscopic separation" has seamless connection with the urban metro and bus system.

The architectural style is "Tophary Design and Villa Vecchia". The whole roof is composed of 11 folded plate steel truss units, the structure of which is in full compliance with their architectural form; each unit has spindle-shaped skylights in the middle high ridges, which can not only use natural light, but more importantly, form a beautiful ridge line on the facades, and match well with both stretching wings. The architecture is quite similar to Wudian-roof with long eaves of the Daming Palace of Tang Dynasty, and represents perfect combination of modern new structural technology and historical context. Due to the folding steel grid structure's mechanical properties perfectly matched the architecture form, making steel for the roof is limited at 68 kg/m^2, the economic benefits are significant.

Many new technologies are used in the Xi'an North Station, such as smart skylight opening control, thermal pressure ventilation, damp energy dissipation and shock reduction, acoustic absorption, noise reduction, and other eco-friendly technologies. It is equipped with accurate, reliable, and practical smart lighting control system; advanced performance-based fire safety design; digital customer service technologies, etc.

西安北站是我国六大铁路客运枢纽站之一，位于西安中轴线北端。旅客流线"上进下出、立体分离"，与城市地铁公交无缝接驳。

建筑寓意为"唐风汉韵，盛世华章"，整个屋顶由11个折板钢网架结构单元体组成，既有中国传统建筑大屋顶的神韵，又是现代新型结构技术的体现，实现历史文脉与现代科技的统一。每个单元体在中间高起的屋脊处开以梭形的天窗，既能利用天然采光，更重要的是立面上形成一道优美的弧形屋脊，与舒展的两翼相得益彰。神似唐代大明宫出檐深远的庑殿顶，是现代新型结构技术与历史文脉的完美结合。由于折板钢网架结构受力特性与建筑形态高度一致，使得屋盖用钢量控制在68kg/m^2，经济效益十分显著。

西安北站大量应用新技术：如利用智能控制开启天窗、热压通风、阻尼消能减振、吸声、降噪等生态环保技术，精确、可靠、实用的智能化照明控制系统，先进的性能化消防设计，数字化的客服技术等。

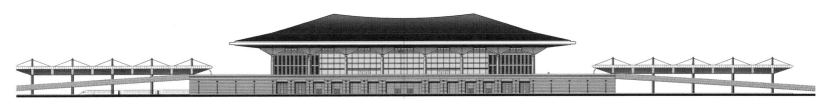

ELEVATION
立面图

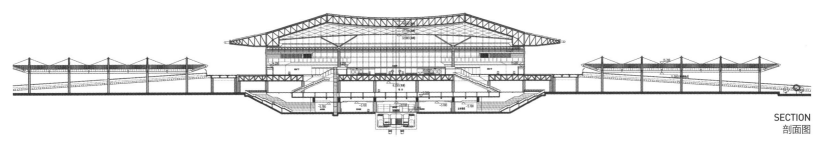

SECTION
剖面图

Tianjin TV Station 天津电视台

Location: Tianjin, China
Architect (Studio): Tianjin Architecture Design Institute
Design: 2006-2012
Construction: 2008-2016
Site Area: 246,000m²
Floor Area: 260,000m²

地点：中国天津市
建筑师（事务所）：天津市建筑设计院
设计时间：2006-2012
建造时间：2008-2016
用地面积：246000m²
建筑面积：260000m²

Tianjin Digital TV Tower is a core architecture of Tianjin TV. The design starts from the internal function and the plane is designed into the Sudoku-style layout, which meets the strict requirement of sound insulation and technology in the studio. Situated at the center of the Sudoku, studios are arranged in spiral escalation on the north side for the convenience of the same floor between each channel and its studio. With Sudoku layout, the purpose of convenience is achieved while all studios are not adjacent to each other, which helps realize sound insulation and saves cost of nearly 30 percent over conventional design, proving to have remarkable economic and social benefits. The space derived from the interwoven aerial studios is regarded as the aerial green atrium, allowing light and air to enter the tower. It is eco-friendly and energy-efficient.

The design philosophy of Media Art Center is "Urban Stage". The architectural volume is combined by two interwoven blocks. One of them is a horizontal block suspending in the air, and another one is a vertical and straight block. A long spacious rotating corridor ascends spirally along with the external audience lounge to guides people to viewing platforms suspended at different levels. The three functions of television theatre, starred hotel and office building are combined organically. For the requirement of television relay function, a 21 x 12 meters large proscenium is designed to meet the demand for 16:9 wide-screen digital television broadcast. An ultra-large transformative performance space is initiated in the Media Art Center domestically with the consideration given to the requirements of traditional theatre.

天津数字电视大厦是天津电视台的核心建筑。设计从内部功能出发，为满足演播厅严格的隔声和工艺要求，将平面设计成九宫格布局，演播厅在九宫格中央和北侧螺旋式上升排布，使各频道办公均能对应本层的演播厅，方便使用的同时又使这些演播厅之间互不相邻，利于隔声，相比常规设计减少约30%的造价，具有显著的经济效益。空中演播厅交错布置产生的衍生空间作为空中绿化中庭，使自然光和空气可以进入大厦内部，达到生态节能的效果。梅地亚艺术中心的设计理念是"城市舞台"，建筑体量由一个悬浮于空中的水平体块与两个垂直挺拔的竖向体块穿插组合而成，一条宽敞的旋转长廊，沿观众休息厅外部盘旋上升，将人流引向悬浮在不同层面的观景平台。建筑中电视剧场、星级酒店和写字楼三大功能有机结合。电视剧场为契合电视转播功能的需求，设计了21mx12m的大型台口，满足16:9宽银幕数字电视转播要求，并首创了国内超大型可变换表演空间，兼顾传统剧场的使用要求。

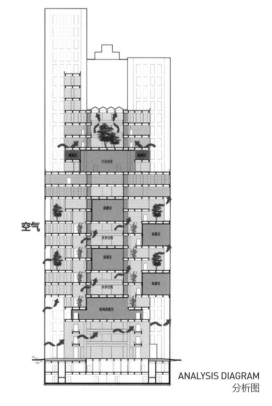

ANALYSIS DIAGRAM
分析图

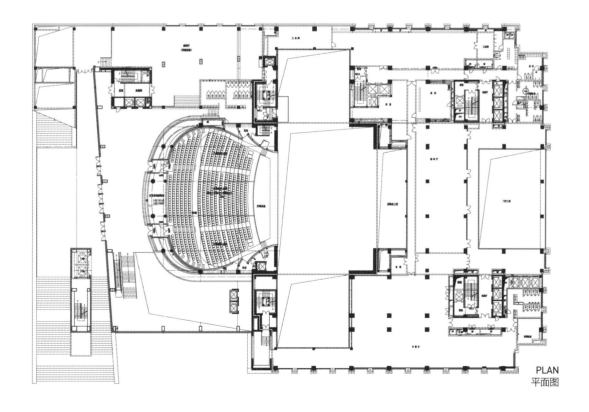

PLAN
平面图

Jixi Museum 绩溪博物馆

Location: Jixi, Anhui Province, China
Architect (Studio): Atelier Li Xinggang, China Architecture Design Group
Design: 2009.11-2010.12
Construction: 2010.12-2013.11
Site Area: 9,500m²
Floor Area: 10,003m²

地点：中国安徽省绩溪县
建筑师（事务所）：中国建筑设计院有限公司李兴钢建筑工作室
设计时间：2009.11-2010.12
建造时间：2010.12-2013.11
用地面积：9500m²
建筑面积：10003m²

Jixi Museum is located in north of Jixi old town in Anhui Province. The entire building is covered under a continuous roof with undulating profile and texture that mimics the mountains and waterways surrounding the county. Once the other buildings in the old town are restored to the traditional Hui style, the museum will fit in even more naturally to the overall townscape.

In order to preserve the existing trees in site as many as possible, multiple courtyards, patios and alleys are introduced into the overall layout of the building. Two waterways are designed along the alleys, eventually converging into the pool in the large courtyard at the main entrance. An introvert courtyard named as "Ming Tang" sits in the southern portion of the building. Directly opposite from the main entrance, there is a group of abstract rockery. A tourism route goes around "Ming Tang" and guides visitors to the "sightseeing platform" at the southeast corner of the building, where they can overlook the building's roof-scape, the courtyards and the mountains in distance.

Triangular steel structural trusses adapt well to the undulating roof. Local building materials, such as stones and clay roof tiles, are used in modern and innovative ways to pay respect to history yet respond to our present times.

绩溪博物馆是一座包括展示空间、4D影院、观众服务、商铺、行政管理和库藏等功能的中小型地方历史文化综合博物馆。

整个建筑覆盖在一个连续的屋面下，起伏的屋面轮廓和肌理仿佛绩溪周边山形水系，又与整个城市形态自然地融为一体。

为尽可能保留用地内的现状树木，建筑设置多个庭院、天井和街巷；沿街巷内部设置东西两条水圳，汇聚于主入口大庭院内的水面；南侧设内向型的前庭——"明堂"，符合徽派民居的布局特征；主入口正对方位设置一组抽象化的"假山"。围绕"明堂"、大门、水面设有对市民开放的立体"观赏流线"，将游客缓缓引至建筑东南角的"观景台"，俯瞰建筑的屋面、庭院和远山。

规律性组合布置的三角屋架单元，适应连续起伏的屋面形态；在适当采用当地传统建筑技术的同时，灵活使用石、瓦等当地常见的建筑材料，并尝试使之呈现出当代感。

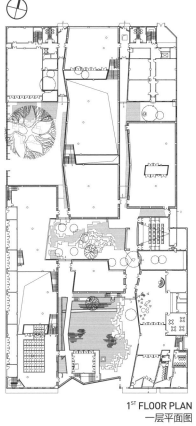

1ST FLOOR PLAN
一层平面图

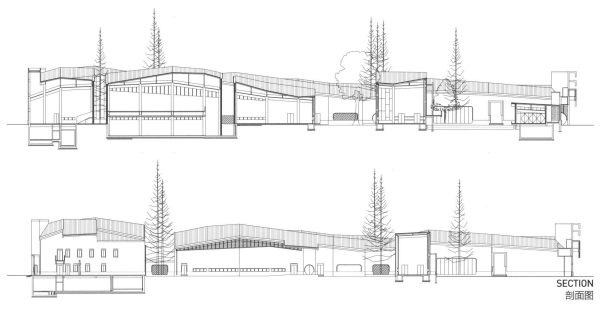

SECTION
剖面图

121

Hongqiao International Airport T1 Renovation and GTC Project

虹桥国际机场 T1 航站楼改造及交通中心工程

Location: Shanghai, China
Architect (Studio): Guo Jianxiang, East China Architectural Design & Research Institute, Arcplus Group PLC
Design: 2012.02-2015.09
Construction: 2014.11-2016.10
Site Area: 103,227m²
Floor Area: 203,746m²

地点：中国上海市
建筑师（事务所）：华建集团华东建筑设计研究总院；郭建祥
设计时间：2012.02-2015.09
建造时间：2014.11-2016.10
用地面积：103227m²
建筑面积：203746m²

Since 1921, the Hongqiao International Airport T1 terminal have witnessed about 100 years' vicissitudes of the city. To promote a comprehensive transformation of the surrounding areas of the airport, the government starts the Hongqiao T1 Renovation Project to improve the passengers' service and build an excellent terminal.
1. Integrated and Diversity: The buildings in terminal area would be made into an integrated terminal complex.
2. Architecture Harmony with Environment: External space design would be integrated with the form design, setting more than 20,000m² concentrated green spaces.
3. Convenient, Comfortable, Cultural and Smart Terminal: The internal functions of the terminal are re-integrated to provide complete and exquisite passenger services. For example, the addition of traffic center functions, the introduction of intelligent and smart technology equipment, and re-planning of concessions, etc.
4. Heritage of Context and Remodeling Space: Re-interpret the existing terminal and absorb its architecture elements, then express them in another modern architectural language.
5. Sustainable Model of Low Carbon Operation: Based on the characteristics of the projects, the architects focus on sustainable design and tailored passive green building strategy.
6. Independent and Integrated Construction: The terminal transformation uses a phased replacement mode, to ensure completed passenger process in each phase.

1921年辟建至今，虹桥国际机场T1航站楼见证了上海近百年的沧桑巨变。为带动机场周边区域的综合改造，实现"脱胎换骨"的转变。虹桥国际机场启动T1航站楼改造项目，整合航站区规划布局，提升整体服务品质，重塑建筑空间形态，打造精品航站楼。
1. 一体综合，功能多样：将航站区建筑群体打造成为一体综合、功能多样的航站楼综合体。
2. 环境建筑，相互映衬：外部空间环境与建筑形态一体化设计，集中设置2万多平方米绿化景观。
3. 便捷舒适，人文智慧：航站楼内部功能重新整合，提供完备、精致的出行服务。比如：增设交通中心功能、引入智能化、信息化设备、商业重新策划等。
4. 传承文脉，重塑空间：通过对原有航站楼建筑的深入解读，汲取既有建筑中的元素，融汇现代设计手法重新演绎。
5. 低碳运行，绿色典范：同样关注绿色节能，立足既有建筑项目特点，量身定制被动式的绿色策略。
6. 易分易合，技术成熟：航站楼改造采用分阶段、置换改造方式，在实施的各阶段，旅客流程都是完整的。

PLAN
平面图

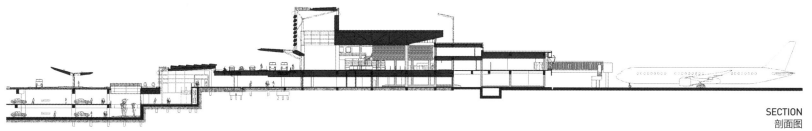

SECTION
剖面图

Power Station of Art 上海当代艺术博物馆

Location: Shanghai, China
Architect (Studio): Zhang Ming, Zhang Zi, Tongji Architectural Design (Group) Co., Ltd
Design: 2011.04-2011.09
Construction: 2011.09-2012.09
Site Area: 19,103m²
Floor Area: 41,000m²

地点：中国上海市
建筑师（事务所）：同济大学建筑设计研究院（集团）有限公司：章明、张姿
设计时间：2011.04-2011.09
建造时间：2011.09-2012.09
用地面积：19103m²
建筑面积：41000m²

It is an art museum within our reach, a spiritual home for sharing artistic experience fairly and also an urban public life platform full of humanistic care. its glorious mission in the industrial age has been finished in the way of historical narrative. It shows the transition from the main workshop of the former Shanghai Nanshi Power Plant to the Urban Future Pavilion of 2010 Shanghai World Expo, now known as the Shanghai Power Station of Art. This six-year arduous design process has witnessed that the previous giant machine for energy output has been transformed into a powerful engine for promoting the development of culture and art.

Limited intervention towards the Nanshi Power Plant gives prominence to the original order and industrial site features of its external structure and internal space to the maximum. At the same time, it deliberately maintains the visible traces of the span of time and space, and gives expression to the unique architectural features of the coexistence of old and new.

The Shanghai Power Station of Art fuses with urban public cultural life with its open and positive attitude, and deliberately obscures the interface between public space and exhibition/showcase space with the scalability of space. It does not only create more opportunities for disrupting the relationship between humans and exhibits in the conventional sense, but also provides the largest possibility for introducing routine situation.

The design illustrates a deep relationship between humans and art by diversified and compound expression of culture, removes the barriers existing in the closed paths of exhibition buildings in the past, and creates a variable atmosphere of exploration.

它是一个触手可及的艺术馆，一个公平分享艺术感受的精神家园，更是一个充满人文关怀的城市公共生活平台。它以一种历史叙事的方式结束了其辉煌的工业时代的使命，经历了从原上海南市发电厂主厂房到2010年上海世博会城市未来馆的转变，继而蜕变为上海当代艺术博物馆。六年的艰辛设计历程见证了一个昔日能源输出的庞大机器如何转变为推动文化与艺术发展的强大引擎。

它对原有南市电厂的有限干预，最大限度地让厂房的外部形态与内部空间的原有秩序和工业遗迹特征得以体现，同时又刻意保持了时空跨度上的明显痕迹，体现新旧共存的特有的建筑特征。

它以开放性与日常性的积极姿态融于城市公共文化生活，以空间的延展性蓄意模糊了公共空间与展陈空间的界定，不仅给颠覆传统意义上人与展品间的关系创造诸多机会，更为日常状态的引入提供最大可能性。

它以多样性与复合性的文化表达诠释人与艺术的深层关系，以漫游的方式打开了以往展览建筑封闭路径的壁垒，开拓出充满变数的弥漫性的探索氛围。

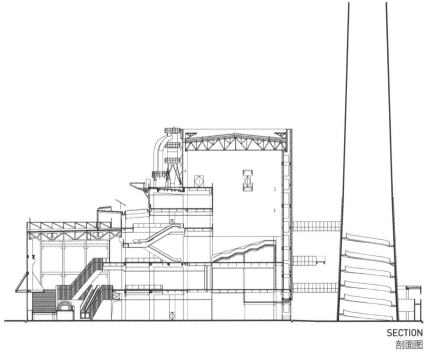
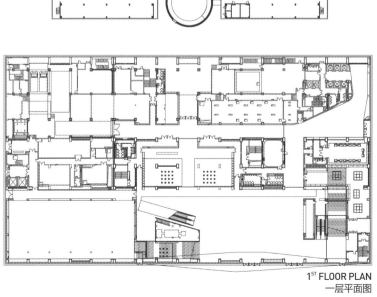

SECTION
剖面图

1ST FLOOR PLAN
一层平面图

Nanjing Yu Garden – A Project to Improve the Surrounding Environment

南京愚园地块保护与整治改善

Location: Nanjing, Jiangsu Province, China
Architect (Studio): Chen Wei, Southeast University School of architecture; Architects & Engineers Co., Ltd. of Southeast University
Design: 2008.04-2014.04
Construction: 2014.04
Site Area: 34,500m²
Floor Area: 3,618m²

地点：中国江苏省南京市
建筑师（事务所）：东南大学建筑学院、东南大学建筑设计研究院有限公司：陈薇
设计时间：2008.04-2014.04
建造时间：2014.04
用地面积：34500m²
建筑面积：3618m²

The Nanjing Yu Garden was architecturally designed as a project to improve the surroundings. It was intended to change chaotic living environment of the shanties that had existed for a long time, through rebuilding the historical site. Meanwhile, it was planned to create a "green lung" for dense residential areas in the vicinities, and attempt to carry forward the city's garden culture.

Respecting history and seeking proven evidence are the first principle to observe in this project. Architects did this by gathering old photos, comparing literature notes, interviewing relevant people, taking adjacent traditional buildings as reference, and rehabilitating existing buildings. As a result, we were able to build a southern section of the garden that looks natural and simple while keeping the exquisite and sophisticated layout of northern section in line with what they used to be in the past.

Scientific positioning and effective utilization are the second principle. Local archaeological excavations gave us a clue to previous water body boundaries. Architects retained and utilized the trees that had grown on the site for a long time, while making up the changed hillsides using rocks, so as to present a picture of natural blending.

Demand adjustment and sustainable development are the third principle. When changing a historical private garden for the purpose of public entertainment, streamlining the garden's layout and readjusting its functionality are a must. Meanwhile, we addressed the needed indoor temperature by taking advantage of the GSHP technology.

To work effectively, architects adhered to on-the-spot communication and collaboration under a reasonably set timetable. As a result, the garden became immediately available to citizens upon its completion, greatly enhancing the availability of an improved environment and citizens' feel of happiness.

作为一个改善整治环境的项目，南京愚园地块建筑设计，目标是通过重建历史上的愚园，改善该地长期以来搭建棚户带来的生活环境脏乱差的问题，同时为密集的居民区提供"绿肺"，也借此传承南京的园林文化。

尊重历史、寻找依据，为第一要则。通过收集照片、比对文献、采访人物、参照周围传统建筑等，修复既有建筑，恢复园南区自然质朴和北区精致密集的历史格局。

科学定位、有效利用，为第二要则。通过局部考古，对淤塞的水体边界进行拟定，对长期养成的树木进行有效利用，对变形的山体以山石进行巧妙补形等，使得大局形态自然混成。

需求调整、持续发展，为第三要则。随着将历史上的私园改善为对市民开放的城市园林的需求，重视流线、调整功能成为必须，同时通过采用地源热泵技术解决了室内环境的使用要求。

同时，设计师坚持现场工作，有效开展专业沟通和合作，时序合理，形成施工完成园即缤纷的景象，大大提高了改善整治环境的有效度和市民使用的幸福指数。

SITE PLAN
总平面图

SECTION
剖面图

Shougang Museum 首钢博物馆

Location: Beijing, China
Architect (Studio): CCTN Architectural Design Co. Ltd.
Design: 2016.08
Construction: 2017.03
Site Area: 72,844m²

地点：中国北京市
建筑师（事务所）：中联筑境建筑设计有限公司
设计时间：2016.08
建造时间：2017.03
用地面积：72844m²

The core parts of the original industrial steelmaking constructions, such as the blast furnace, the hot-blast stove, the gravitational dust collector and the dry dust collectors, are reserved. The control room to the west used to shelter the furnace is opened up to allow the steelmaking constructions to echo with the Xiu Lake and Mount Shijin.

The three pavilions covered under a green slope contains the parts of the museum that are easiest to rent out and share, including a lecture hall, a temporary exhibition space, a restaurant/café, which are available to more people by their externalization. The small-sized buildings adjacent water serves as a smooth transition between the Lake and the blast furnaces.

The circulation starts from the embankment of the Xiu Lake, moves into it along the Shougang Memorial Wall until the underwater garden, with the furnace echoing from behind. It passes through an underwater corridor, goes back to a blast furnace, and then see blast furnaces from a 9.7 meter cast house platform, a 13.6 meter visit platform, a 40 meter cover platform until a 72 meter burner platform, allowing people to take in everything in the entire blast furnace ironmaking process at a glance as the elevations increase. It also leaves deep impression of glory over the years in people's mind by means of an extremely industrial and extremely natural dialogue.

设计谨慎保留了原首钢的明星高炉——三高炉的主体高炉部分、热风炉、重力除尘器和干法除尘器等核心工业构筑物，打开原有在西侧遮挡炉体的主控室让三高炉构筑物群和西侧的秀池、石景山建构了清晰的对话关系。

绿坡覆盖下的三馆抽取了博物馆中最容易对外出租和共享使用的功能（报告厅、临时展厅、餐厅和咖啡厅）外置，使其服务周边更大的园区人群。以活泼的小尺度滨水建筑，柔软地缝合了秀池和高炉的边界。

设计动线先由保留的秀池柳堤深入湖面，沿清水混凝土的首钢功勋墙拾级而下慢慢没入池中，在水下展厅圆形静水院回望高炉。穿过水下廊道回到高炉内部，依次看到高炉从9.7m出铁场平台、13.6m参观平台到40m罩棚平台一直到72m炉头平台，令人在标高攀升的步移景异中一览整个高炉炼铁的全部工艺流程。更让人在极度工业和极度自然的对话中，铭记曾经的岁月荣光。

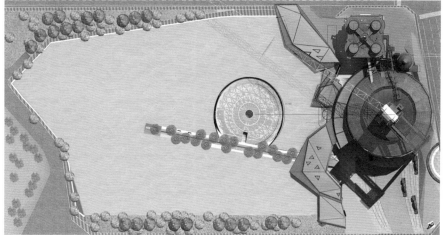

SITE PLAN
总平面图

SECTION
剖面图

Relics Park for the Coal Dock of Xiaguan Power Plant

南京下关电厂运煤码头遗址公园

Location: Nanjing, Jiangsu Province, China
Architect (Studio): East China Architectural Design & Research Institute, Arcplus Group PLC
Design: 2012-2013
Construction: 2013-2014
Site Area: 9,200m²
Floor Area: 3,922m²

地点：中国江苏省南京市
建筑师（事务所）：华建集团华东建筑设计研究总院
设计时间：2012-2013
建造时间：2013-2014
用地面积：9200m²
建筑面积：3922m²

The site was the Coal Dock of Nanjing Xiaguan Power Plant that nearly had one century history. After the Plant moved away, the dock is the only venue to be reused as the Relics Park which will be open to the public. Some existing memorable records about Xiaguan Power Plant will be displayed in the exhibition hall. And people can drink a cup of coffee or have meal while enjoying the river views in the restaurant that is decorated in industrial flavor. Since the Old Dock is a precious industrial heritage and also a valuable historical memorial place, the design should combine historic culture with mechanical aesthetics, use architectural materials to convey the feeling of solemnity.

Adjacent to the project site, Red House, decorated with red brick wall, blue tiled spire, is in typical style of Nanjing during Republic of China. It is important historical remains. Buildings around it are required to be visually harmonious in urban planning. So the façade of reserved buildings were remade by bricks and the added parts by glass curtain wall to make a distinction between old and new. The integrated landscape design consists of industrial site design, landscape design, ecological elements design. Fusion of industrial relics and modern landscape elements: using local materials to reflect industrial relics and modern landscape elements in the site, forms a new type of landscape. Through refining and abstracting industrial cultures in the site to tell the story of history, culture and emotion in the site. Taking industry and wharf as theme, use experiential art of landscape design to restore history, and to inherit spirit.

南京下关电厂煤码头有近一个世纪的历史。该厂搬迁后，码头是唯一可重复使用的场所作为遗址公园向公众开放。有关电厂的历史资料将在展览馆展示，人们还可以在充满工业风的餐厅中喝一杯咖啡。老码头是一处宝贵的工业遗产和历史纪念场所，设计需要将人文历史与机械美学相结合，用建筑材料来传达庄重的感觉。

作为重要的历史街区的一部分，厂区与周边建筑需要实现视觉上的统一。因此建筑采用红色砖墙，蓝瓦尖顶，这正是民国时期南京的典型建筑风格。改造时增加的金属构件和玻璃幕墙则产生了新老碰撞。景观设计理念融合工业场地设计、景观设计、生态元素设计。利用当地材料反映工业遗址和现代景观元素，形成一种新型景观。通过提炼现场的工业文化，然后运用产业元素的组合，在整个脉络中形成文化线索，讲述场所的历史、文化和情感；以工业与码头为主题，运用景观设计的体验艺术，还原历史，传承精神。

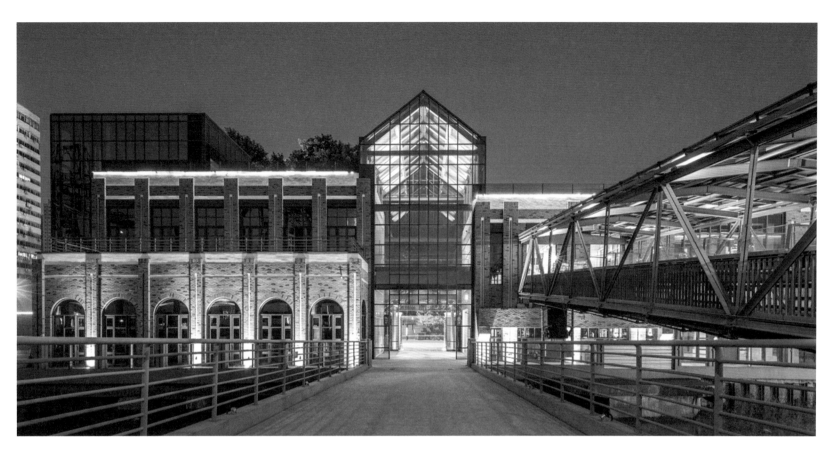

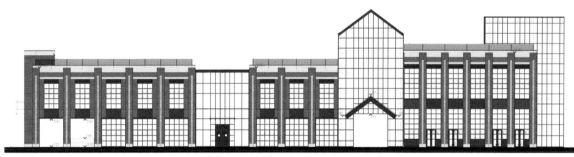

EAST ELEVATION
东立面图

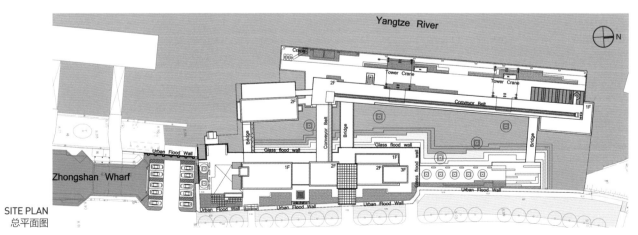

SITE PLAN
总平面图

Xi'an South-Gate Plaza Improvement Project

西安南门广场
综合提升改造项目

Location: Xi'an, Shaanxi Province, China
Architect (Studio): China Northwest Architecture Design & Research Institute Co., Ltd.
Design: 2011.01-2013.09
Construction: 2013.09-2014.09
Site Area: 82,879m²
Floor Area: 69,229m²

地点：中国陕西省西安市
建筑师（事务所）：中国建筑西北设计研究院有限公司
设计时间：2011.01-2013.09
建造时间：2013.09-2014.09
用地面积：82879m²
建筑面积：69229m²

Xi'an South-Gate Plaza Improvement Project, is designed to protect the cultural relics, beautify the environment and rebuild historical features. We added the infrastructure, integrated the Plaza's landscape system, and improved the lines of traffic flow both inside and outside. The project land is naturally divided into several plots:

The Plaza outside South Gate was mainly for the construction of underground parking. square and landscape were used in its landscape design to reshape the image of urban living room and to meet the needs of theatrical performance activities. Besides, 2 E-W underground channels across the Plaza were built to connect MiaoYuan, SongYuan and the central square.

MiaoYuan\SongYuan plots are separated by main plaza as two wings. A series of Sunken courtyards\streets shaped space.

The west\east Citywall Park block next to the south of Xi'an city wall. We designed a earth sheltered architecture on each part. The building adopted minimalism modern style to express the respect of cultural relics.

Plot South Gate: The south extension of underground pedestrian channel for South Street was built to introduce pedestrians from underground to the South Gate area and separate people and traffic.

西安南门广场综合提升改造项目，设计旨在提升区域环境品质、重塑古建历史风貌的前提下，补充完善景区配套设施；整合广场景观系统，完善城墙内外各方向车行、步行、轨道交通流线。项目用地自然地划分为若干地块：
南门外广场主要是建设地下停车场。景观设计以广场、绿化重塑城市客厅形象，并满足文艺演出活动需求。另修建两条东西向横穿广场的地下通道，联系起松园、苗园及中央广场。
苗园、松园地块分别位于护城河南岸东、西两翼，建筑以下沉式广场的形式匍匐于场地中，地面各布置有一组庭院式建筑。建筑采用现代坡屋顶的形式，与南门城楼、场地内的保留建筑相协调。
紧挨城墙南侧的东、西环城公园地块，设计以极简主义原则各加建一处幕墙饰面的覆土建筑，掩映于树林之中，与城墙形成对比性协调。

南门里地块：修建南大街人行地下通道南延段，将行人从地下引入南门景区，实现人车立体分流。

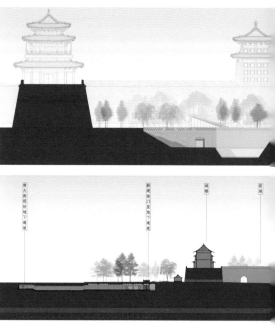

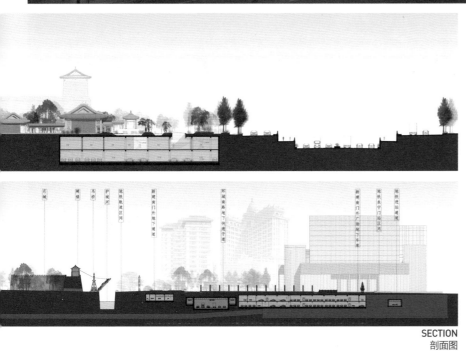

SECTION
剖面图

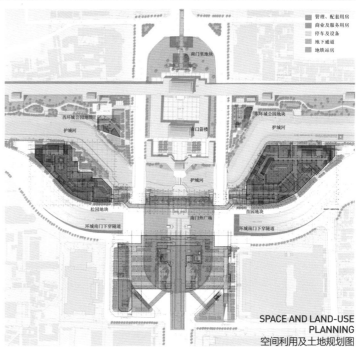

SPACE AND LAND-USE PLANNING
空间利用及土地规划图

Historic Block Space Regeneration -Seclusive Jiangnan Boutique Hotel

历史街区空间再生
——隐居江南精品酒店

Location: Hangzhou, Zhejiang Province, China
Architect (Studio): Zhejiang Greenton Architectural Design Co., Ltd.
Design: 2016.01-2016.04
Construction: 2016
Site Area: 1,237m²
Floor Area: 2,816m²

地点：中国浙江省杭州市
建筑师（事务所）：浙江绿城建筑设计有限公司
设计时间：2016.01-2016.04
建造时间：2016
用地面积：1237m²
建筑面积：2816m²

Seclusive Jiangnan Boutique Hotel is located in Dadou Road Historic District, Hangzhou, and adjacent to Beijing-Hangzhou Grand Canal. Before the renovation, the existing structures were two dilapidated affordable housings. As the general contractor, we found a balance between two pairs of seemingly incompatible paradoxes "preservation" and "demolition" as well as between "inheritage" and "innovation", regenerated the historical district through design.

The project aims to convey people the idea that "spatial quality is the soul for residential experience". The original form of buildings is kept, but building space, size and line of flow are reorganized. Building 188# is I-shaped and faces the canal while Building 190# is L-shaped, tranquil and elegant adjacent. Designers insert a glass box as hotel lobby and main entrance that connects the two separate buildings. In this way, it forms a courtyard as a 3-side enclosed space. The further design demolished the excess volume so that to keep the façades clean and straight.

In the facade design, the side facing a historical district is introverted and its traditional texture is touched up with black bricks. Modular bay windows are used to increase the available area. Equipment is hidden in metal mesh. The splicing of old and new materials realizes the isomorphism of modern and traditional streets and minimizes the intervention in traditional streets. The facades for Floor 2 and Floor 3 along the canal are lattice brick walls. The original texture is touched up and the building volume is integrated. Floor standing windows are staggered on both floors. The whole building looks both tough and soft.

隐居江南位于杭州大兜路历史文化街区，紧邻京杭大运河，项目改造前为两幢现已破败的四层安置房建筑。项目团队在不到一年的时间里，平衡"保留"与"拆除"、"延续"与"创新"这两对看似相悖的矛盾体，通过设计让历史街区中的空间进行再生。

本项目旨在无形中向人们传递"空间才是居住体验的灵魂"的理念。设计保留原有建筑的基本形态，重新梳理建筑的空间、体量、流线。188号建筑呈"I"形，直面运河，190号建筑呈"L"形，安静雅致。在两单体间置入玻璃盒子作酒店大堂，连接底层空间，对外为主入口。酒店外部增加围墙，三面围合，界定空间，自然而成的中庭内向收敛。形体设计从"加、减"两个角度切入。先做"减法"，拆除多余建筑体量，再做"加法"，填平立面。

在立面设计上，面向历史街区的一侧相对内敛，用青砖修补传统肌理，用模数化的飘窗增加使用面积，用金属网隐藏设备，新旧材质的拼接，完成现代与传统街巷语言的同构，最大程度地减少对传统街巷的干预。沿运河二三两层，立面使用整片的花格砖墙，修补原始肌理，整合建筑体量，落地窗上下两层交错，建筑整体呈现硬朗与柔软的双重特质。

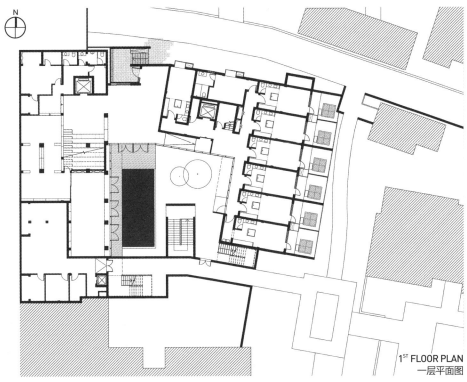
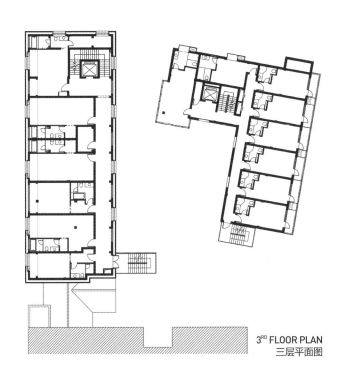

1ST FLOOR PLAN
一层平面图

3RD FLOOR PLAN
三层平面图

TENIO Green Design Center 天友绿色设计中心

Location: Tianjin, China
Architect (Studio): Tenio (Tianjin) Architecture and Engineering Co., Ltd
Design: 2011.11-2012.04
Construction: 2012.04-2012.12
Site Area: 3,215m²
Floor Area: 5,756m²

地点：中国天津市
建筑师（事务所）：天友（天津）建筑设计股份有限公司
设计时间：2011.11-2012.04
建造时间：2012.04-2012.12
用地面积：3215m²
建筑面积：5756m²

Tenio Green Design Center is located in a reconstructed multistory factory building. By virtue of problem-oriented technology integration, and using some innovative and experimental technologies as well as some technologies appropriate to North China areas, it achieved the goal of low-cost reconstruction with ultra-low energy consumption. The green technologies used include adjustable climate cavity, moveable heat insulating wall, polycarbonate super insulating curtain wall, vertical greening, roof agriculture, modular terrestrial heat pump, separate temperature and humidity control air-conditioner, floor radiation heat/cold supply, free cold supply, and so on. The total energy consumption of this building during operation is 40.17kW·h/m²·a, and energy consumption of air conditioning and heating is as low as 19.6kW·h/m²·a, so the objective of ultra-low energy consumption is achieved. This project won ARCASIA Gold Medal (2014).

天友绿色设计中心是一座既有多层厂房的绿色化改造，借助问题导向的技术集成方法，设计选择中国北方适宜技术，同时创新性地应用一些实验性技术，实现了超低能耗的低成本改造目标。绿色技术体系包括可调节气候腔体、活动隔热墙、聚碳酸酯超级保温幕墙、垂直绿化、屋顶农业、模块式地源热泵、温湿度独立控制空调、地板辐射供冷供热、免费供冷等。建筑运营阶段的总能耗为40.17kW·h/m²·a，空调采暖能耗仅为19.6kW·h/m²·a，达到了超低能耗的目标。项目荣获2014年亚洲建协可持续建筑金奖。

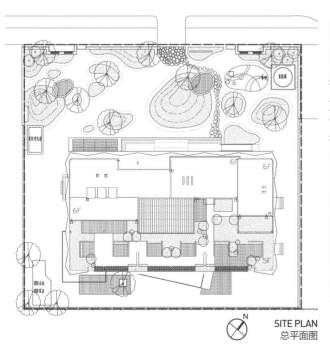

SITE PLAN
总平面图

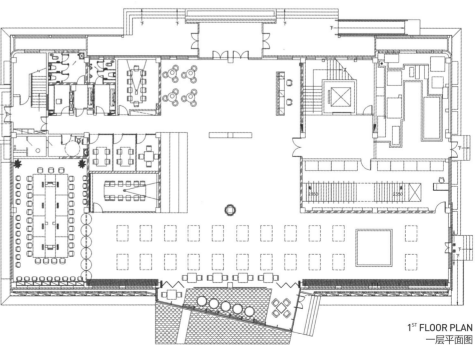

1ST FLOOR PLAN
一层平面图

Quadrangle Renovation, Caochang Area, Beijing (Yard 8,19,36,41)

北京前门草厂片区四合院改造

Location: Bejing, China
Architect (Studio): Beijing Institute of Architectural Design Co., Ltd
Design: 2015.03-2015.10
Construction: 2015.10-2016.08
Site Area: 4,360m²
Floor Area: 3,160m²

地点：中国北京市
建筑师（事务所）：北京市建筑设计研究院有限公司
设计时间：2015.03-2015.10
建造时间：2015.10-2016.08
用地面积：4360m²
建筑面积：3160m²

Caochang Area is one of the key protected areas among the 25 historical protection areas of Beijing. It ranges from West Xinglong Street to the north and Caochang Shitiao to the east, and borders Caochang Santiao in the west and North Lucaoyuan Hutong, Xuejiawan Hutong in the south.

Caochang has formed special hutong characteristics over the years. Most houses and courtyards in Caochang Area had been rehabilitated and renovated in 2008, contributing to the improvement of living environment in this area. Whereas inadequacies remained, for instance, municipal infrastructures are insufficient to satisfy the demand for modern living. As time passed by, many of original residents moved out and many courtyards were left vacant while the rest of the residents chose to stay and pass the anecdotes they had experienced to the next generations.

In March 2015, designers retrofited the courtyards within Caochang Area. The courtyards approved for retrofit weremostly within Caochang Sitiao and Caochang Wutiao. They were totally 26 courtyards, dwelled or vacant. Prior to the retrofit, the designer teams had made intensive investigations and were deeply impressed by abundant quadrangle courtyards as well as different-class residents, such as neighborhood of dignity, ordinary civilians, and royal descendents. Therefore, the designers thought that protection of original residents as well as of architectures are equally important.

The retrofit included retrofit of courtyards buildings, enhancementof hutong environment and retrofit of municipal infrastructure. The courtyards and architectures are meant to be modified into a complete set of units by equipping every dwelling unit with bedroom, living room, dining room, kitchen, restrooms. Landscape enhancement aimed to make sure residents enjoy good municipal environment and the municipal infrastructure retrofit shall guarantee efficient use of buildings. On such basis, the designer teams prepared renovation guidelines to guarantee an overall control of courtyard retrofit.

草厂片区是北京旧城25片历史文化保护区中重点保护区之一，具体范围北至西兴隆街，东至草厂十条，西临草厂三条，南至北芦草园胡同、薛家湾胡同。

在岁月的更替中胡同形成了独特的肌理，延续发展至今。目前草厂片区的大部分房屋于2008年进行了修缮和翻建，一定程度上改善了片区的居住环境，但仍有一些不足之处，如市政设施不齐全，难以满足现代生活的需要等。随着时间推移，很多原住户也陆续搬离了本片区，腾出了很多空院，也有大量居民选择继续生活在这里，他们在这里相传历史的轶事，见证岁月的变迁。

2015年3月，对前门草厂片区进行院落升级改造，此批改造的院落集中在草厂四条及五条，共26个院落，分为有人居住的带户院及空置的空院两种。在改造之前，设计团队对本片区进行了多次深入的调研，过程中给设计师留下深刻印象的不单是片区中丰富的四合院，更是内部多样的居民。这里有曾经达官显贵的邻居，有普通市民，也有皇族后裔，阶层丰富多彩，坊间流传的故事也很多。因此设计师提出，本片区的改造中原住户的保护与建筑的保护同样重要。

本次改善主要分为：院落建筑改造、胡同环境提升及市政设施改造三个方面展开，建筑改造力求将内部的每个居住单元都改造成有卧室、客厅、餐厅、厨房、卫生间的成套户型，景观提升要能确保居民可以享有良好的市容环境，市政设施改造需保障建筑的有效使用。在此基础上，制定了院落修缮的导则，以期对院落改造有整体的控制。

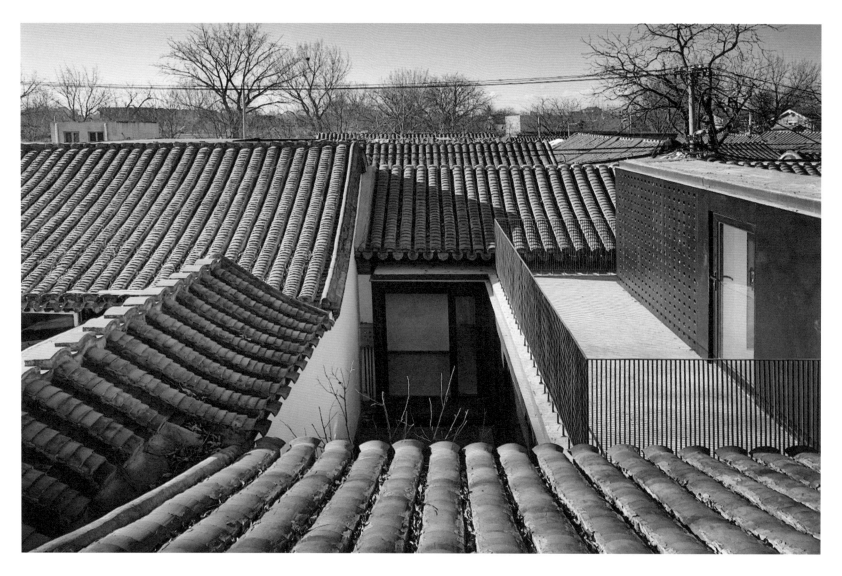

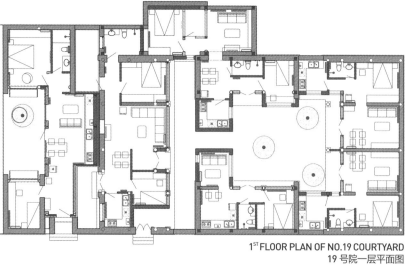

1ST FLOOR PLAN OF NO.19 COURTYARD
19 号院一层平面图

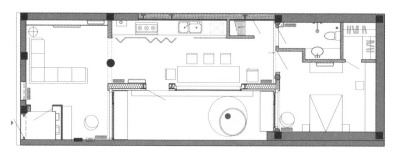

1ST FLOOR PLAN OF NO.8 COURTYARD
8 号院一层平面图

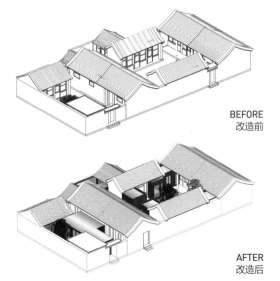

BEFORE
改造前

AFTER
改造后

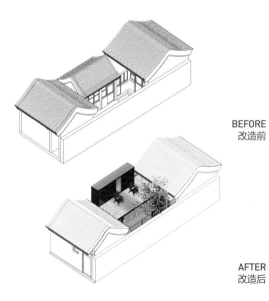

BEFORE
改造前

AFTER
改造后

Tianning NO.1 Culture and Creative Industrial Prak

天宁一号 文化科技 创新园

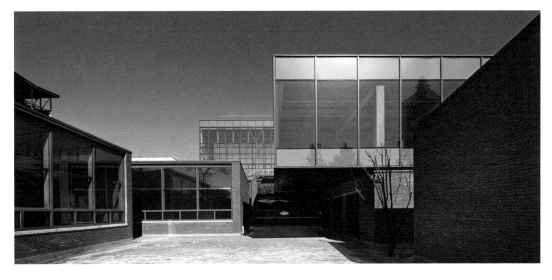

Location: Bejing, China
Architect (Studio): UFO, Beijing Institute of Architectural Design Co., Ltd.
Design: 2015-2016
Construction: 2016
Site Area: 7,900m²
Floor Area: 80,000m²

地点：中国北京市
建筑师（事务所）：北京市建筑设计研究院有限公司方案创作工作室
设计时间：2015-2016
建造时间：2016
用地面积：7900m²
建筑面积：80000m²

Surrounded by the historical relics of the capital of both Jin Dynasty and Liao Dynasty, the former site of Beijing No. 2 Thermal Power Plant is boarded by the Tianning Temple Pagoda of the Northern Wei Dynasty to the east, the earliest exiting historic building in Beijing, by the White Cloud Temple and Jiqiu Relics Park to the north. At the earliest, it used to be a side hall of the Tianning Temple Pagoda and the venue of Emperor's Temple during the Liao Dynasty. It is not only the earliest cradle of city but also the historic and cultural origin of coordinates in Beijing.
1. Retain the original structure of the plant and its industrial architectural style;
2. Keep the existing architectural scale and the scale of greenbelt.
Retrofit Strategies:
1. We should enhance its connection with its surrounding historic and cultural buildings in terms of sight lines and moving lines and provide the park with richer cultural elements;
2. The city boundary should be opened up so as to enhance external communication and inject the city's vigor into the Park;
3. Internal recycling conditions should be improved to make internal communication more convenient;
4. Various kinds of office occupancies should be integrated to enhance uniformity and wholeness;
5. The hierarchy of courtyard space and urban texture should be improved; a variety of extensive exchange and display and landscape space should be provided.
Through the above strategies, the garden-styled urban characteristic of "one core zone and three parks" is highlighted on the basis of the original plant, to form a cultural and technological theme park that is rich in connotation, energetic, organic, and highly efficient. The characteristics of "one core zone and three parks" represents the highlight of public service platform in the park and emphasizes the connection and interaction among all functional modules, as well as the core concept of a green urban park where resources and space can be shared, so that the features of the former site of the thermal power plant is not only retained, but also an epoch-making innovative factory is born.

项目位于北京第二热电厂旧址，周边为金代和辽代都城遗址，东侧紧邻北京现存最早的古建筑北魏天宁寺塔，北侧为白云观和蓟丘遗址公园。基地本身最早为天宁寺配殿和辽代历代帝王庙所在地，它是北京的最早城市发源地，也是北京历史文化的坐标原点。
1. 保留原有厂区格局和工业建筑风貌；
2. 保持现有建筑规模和绿地规模。
改造策略：
1. 加强与周边历史文化建筑之间的视线和动线联系，为园区提供更加丰富的文化元素；
2. 开放园区边界，增强对外联系，为园区注入城市活力；
3. 改善内部循环条件，增加内部交流的方便性；
4. 整合各类办公用房，加强统一性和整体性；
5. 增加院落空间层次，改善城市肌理，提供丰富多样的交流展示与景观空间。

通过以上策略，在原有厂区基础上突出了"一核三园"的园林式城市特色，形成了内涵丰富、活力充沛、有机高效的文化科技主题园区；"一核三园"代表了园区突出公共服务平台，强调各功能板块互联互通，资源和空间共享的绿色城市型园区核心理念，既保持了原有老旧厂区的特色，又产生了一座划时代意义的创新工场。

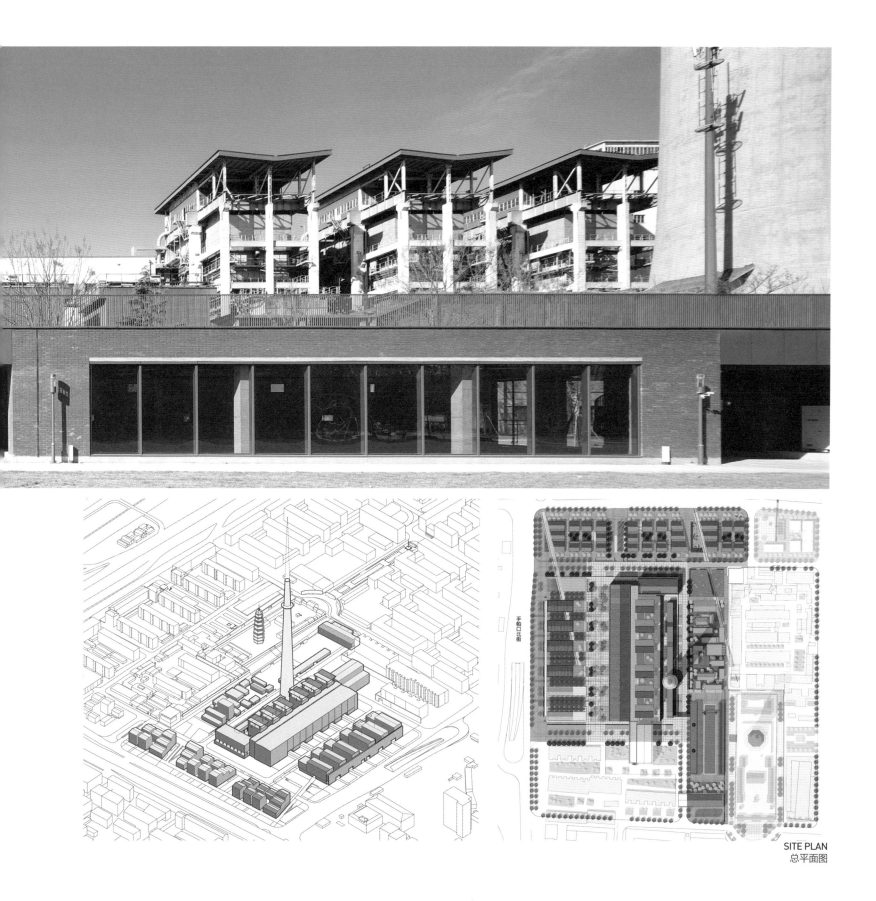

SITE PLAN
总平面图

161

Comprehensive Renovation Project of Zhongshan Road (from Jianghan Road to Yiyuan Road)

中山大道综合改造工程（江汉路至一元路）

Location: Wuhan, Hubei Province, China
Architect (Studio): Xiao Wei, Qi Wei, Wang Yiying, Li Yuting, Zhang Xi, Zhang Wei, Zhou Xuan, Shang Yuting, CITIC General Institute of Architectural Design and Research Co., Ltd.
Design: 2015.04
Construction: 2016.05
Floor Area: 213,350m²

地点：中国湖北省武汉市
建筑师（事务所）：中信建筑设计研究总院有限公司；肖伟、齐蔚、王一莹、李宇庭、张曦、张伟、周璇、尚宇婷
设计时间：2015.04
建造时间：2016.05
建筑面积：213350m²

Zhongshan Road, firstly built in 1906, is one of the key avenues of Old Hankou with irreplaceable functions of commerce and transportation, and listed as one of China's first national historical and cultural districts. In order to develop it into an avenue of culture, tourism and landscape, a series of special projects, including urban space, building facades, landscape and public works, were carried out in 2015. The project covers a distance of 4.7 kilometers starting from Yiyuan Road to Wusheng Road, among which the 1.7 kilometers from Yiyuan Road to Jianghan Road runs through the British, Russian, French and German concessions and holds the historical and cultural landscape of Wuhan in different periods. About 80 buildings are built in this section, including 7 cultural relics, 19 municipality's excellent historical buildings and 6 unmovable cultural relics.

"Manifest historic landscape, reflect Wuhan features, and inherit urban foundation"
1. Keep the relatively high level of building density, with historic streets and sites protected and renovated under the principle of preserving as many valuable relics at different periods as possible.
2. Maintain cultural value and realize sustainable utilization.
3. Distinguish public space from private space, selectively remove those residential buildings which are damaged seriously or conflict with historical features, rehabilitate the tissue of district, establish a public space system, rebuild social place, and create an urban micro-landscape.
4. Optimize the transportation system, by means of reducing the space for vehicles, widening the space for pedestrians and emphasizing humanization.
5. Propose precise protection measures for facades according to construction time, appearance features and current status, keep rehabilitating featured buildings, and improve the protection list.
6. Improve the living environment, put municipal pipelines underground, renew the lightning and identification systems, and undertake the energy-saving renovation item by item.
7. improve the list of major buildings to be protected, to provide support for further setting up and implementing the mechanism of maintaining cultural and historic relics.

中山大道，始建于1906年，是老汉口商业交通性干道，中国首批国家级历史文化街区。2015年根据"文化旅游景观大道"发展定位，全面开展城市空间、建筑立面、园林景观、市政工程等专项工作。总体工程从一元路起至武胜路共4.7km，其中一元路至江汉路段全长1.7km，跨越原英、俄、法、德四大租界区，沉淀了武汉不同时期的城市历史人文景观。该段涉及文物7处，市优秀历史建筑19处，不可移动文物6处，共80余栋建筑。
"彰历史风貌，显武汉特色，续城市之根"
1. 整体保持片区相对高密度，重点保护和改造蕴含城市历史的街道和场所，尽可能多地保留各个时期有价值的遗存。
2. 维护文化价值，实现可持续利用。
3. 明确公共与私人空间界限，选择性拆除破损严重与历史风貌冲突的民居，修复街区肌理，建立公共空间系统，恢复交往场所，打造城市微景观。
4. 实施"公交街道"改造，压缩车道规模，保障步行交通，注重人性尺度。
5. 根据保护建筑年代、风貌特征及现状逐栋确定精准化立面保护措施；继续修缮和升级风貌建筑，完善保护名录。
6. 改善人居环境，实施市政管线入地，亮化及标识系统更新，实行分项节能改造。
7. 完善重点保护清单，为进一步制定并落实文物古迹保养制度提供了研究支持。

Hexing Warehouse Renovation, 2010 Shanghai Expo Park, China

中国 2010 年
上海世博会
和兴仓库改造

Location: Shanghai, China
Architect (Studio): Huang Qiuping, Huang Wei, East China Architectural Design & Research Institute, Arcplus Group PLC
Design: 2008.10-2009.06
Construction: 2009.06-2010.04
Site Area: 16,521m²
Floor Area: 2,241m²

地点：中国上海市
建筑师（事务所）：华建集团华东建筑设计研究总院：黄秋平、黄巍
设计时间：2008.10-2009.06
建造时间：2009.06-2010.04
用地面积：16521m²
建筑面积：2241m²

Hexing Warehouse witnesses the emerging of first private steel enterprises in China. It was built in 1930's and featured in gentle inclined scissors type stairs, reinforced concrete beams with variable cross-sections, flat-plate floor structural system and memorial archway style facade towards Huangpu River. As a representative of modern industries in Shanghai and a typical dock warehouse, the building is kept during Shanghai Expo park construction, as homage to history, culture and heritage buildings.

The renovation of the Warehouse respects "originality" and it reproduces old materials and construct craft. To reinforce structure and increase seismic bear capacity in accordance with present code requirements, the designers, after thorough comparisons and all around analysis, decided to embed steel cross braces in the most vulnerable places and add horizontal bars to link the existing construction and the new.

The designers insert a new rectangular volume into the Warehouse to incorporate new functions. The volume is finished with aluminium rods of simple vertical wood textures to highlight the contrast between old and new. The renovation has reassured general concern of keeping the building, and since then turns over a new page in the Warehouse's history.

和兴仓库作为中国第一家民营钢铁企业的历史见证物具有一定的历史文化价值，它的保留既反映了上海近代工业的文明，又反映出上海码头仓库的历史建筑风貌，并表明了上海世博会对城市历史和特有历史建筑的尊重。其面向黄浦江的牌坊式立面、平缓的交叉楼梯、钢筋混凝土变截面交叉梁及无梁楼盖结构体系具有20世纪30年代老仓库的独特风格，具有一定的保留价值。

和兴仓库的改造遵循"原真性"的原则，从而可反映出当时的建筑材料、建造工艺。据此，结构方案在多方案比较、精确分析后决定采用全新的加固方法以满足抗震规范的设计要求，即在受力最薄弱的部位加钢人字撑来承受部分地震力。新老建筑之间在传力途径上加设水平撑杆，以承担其余部分的地震水平力。

和兴仓库中间加入长方形筒状体量，以满足新的功能要求，新体量以简洁的竖向木纹铝条装饰，以对比的方式衬托出老建筑的历史风貌。

和兴仓库的改造建设，使一个饱受保留和拆除争议中的建筑暂时告别喧嚣，从而翻开它新的历史一页。

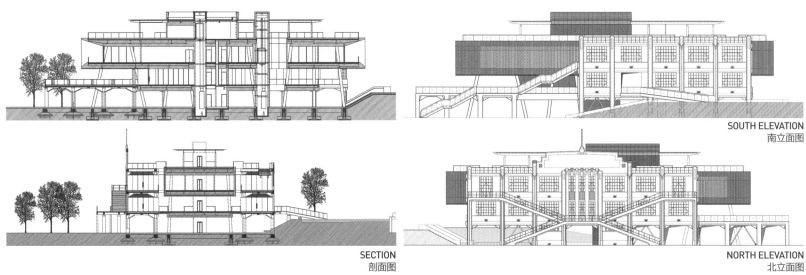

SECTION
剖面图

SOUTH ELEVATION
南立面图

NORTH ELEVATION
北立面图

Preservation and Reparation Project of the Capital Cinema

首都电影院

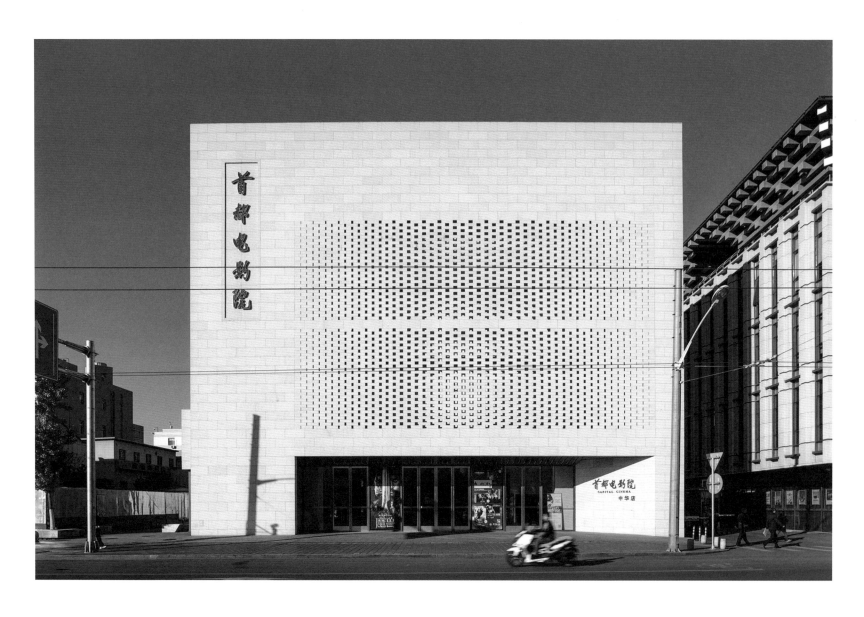

Location: Beijing, China
Architect (Studio): China IPPR International Engineering Co., Ltd.
Design: 2014.07-2015.10
Construction: 2016.02
Site Area: 1,304m²
Floor Area: 2,232m²

地点：中国北京市
建筑师（事务所）：中国中元国际工程有限公司
设计时间：2014.07-2015.10
建造时间：2016.02
用地面积：1304m²
建筑面积：2232m²

Being the central point in Tianqiao district, the Capital Cinema has undergone many repairs and renovations from 1987 to 2014. The major principle of this renovation is to preserve the original structure while transforming interior spaces. Designers cleared up the interior of the second and third floor and convert it into IMAX theatre. We reformed the first and the underground floor, and set two small movie halls, and arranged scattered stores in the public space.

In facade design, designers use the main decorative elements of surrounding buildings to ensure a uniform style of the central axis. We use beige granite and they are partially scattered to form a hollow feeling in skin texture. This facade style coordinates with the Tianqiao arts center, and represents strong characteristics of the time and the region.

作为天桥地区的重要节点，首都电影院在1987年到2014年间，历经变迁。在建筑的改造中，设计师保留了原始的结构，将二层和三层的内部空间整体拆除，建造一座巨幕影厅；对首层及地下进行改造，建造两座影厅，并结合公共空间布置商业。

在建筑立面的设计中，设计师选择周边建筑的立面元素，以保证整个中轴线建筑风格的统一。周身装饰淡黄色石材，局部采用石材错位砌筑的形式，形成镂空的表皮肌理，既与天桥艺术中心表皮肌理相契合，又体现一定的时代特征，延续了南中轴线建筑风格。

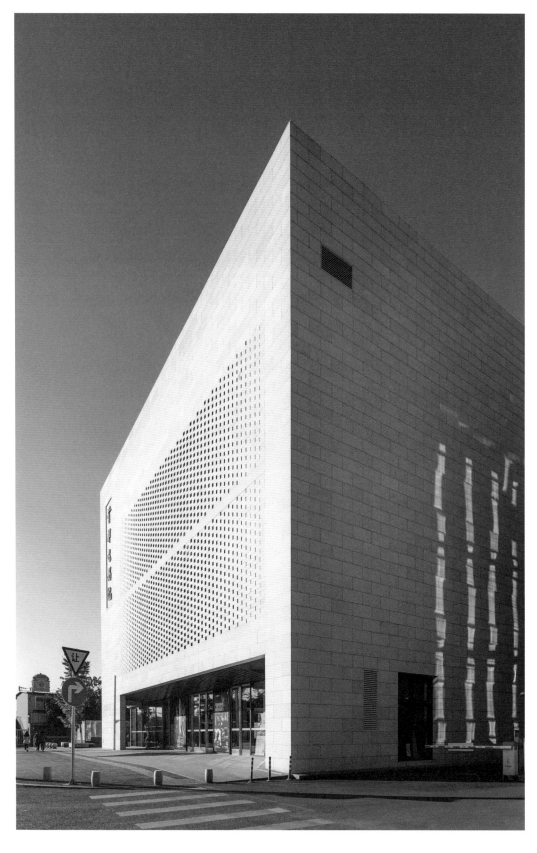
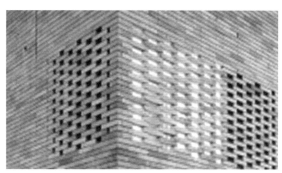

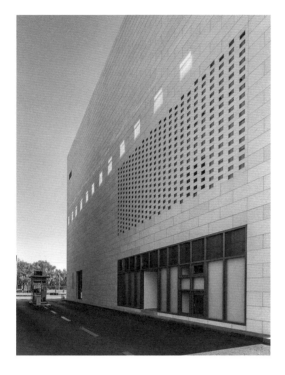

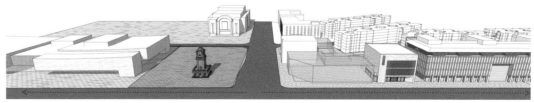

Tianqiao Art Centre　　Civic Square　　Tetrahedral Clock　　Beiwei Road　　Second-phase Project　The Capital Cinema　　Tianqiao Art Office

ANALYSIS DIAGRAM
分析图

173

Landscape and Related Facilities of The 9th China (Beijing) International Garden Expo Park

第九届中国（北京）国际园林博览会园区绿化景观及相关设施建设项目

Location: Beijing, China
Architect (Studio): Design Department EA4, Beijing Institute of Architectural Design Co., Ltd
Construction: 2011.01-2013.04
Site Area: 2,670,000m²

地点：中国北京市
建筑师（事务所）：北京市建筑设计研究院有限公司 EA4 设计所
建造时间：2011.01-2013.04
用地面积：2670000m²

The 9th China (Beijing) International Garden Expo has displayed Chinese and international garden history, culture and artistic accomplishments. It provides an interactive platform of garden culture and technology for the public. Through ecological restoration, a former waste dumping site is now transformed into a popular urban park, a great achievement in environmental development.

By connecting the city with Yongding waterfront, the Beijing Garden Expo Site is a catalyst for the Capital city's south "Waterfront Economic Belt" development plan. It accelerated the construction of roads, subway, high speed railway and electrical infrastructures. It promotes traffic construction in southwest area of Beijing and lays foundation of sustainable development for the region. The Expo also pushes forward the development of Yongding River Ecological Corridor. The overall improvement of the environmental conditions has benefited the citizens in southern area.

第九届中国（北京）国际园林博览会通过全面展示中国及世界园林历史、文化和艺术成就，为广大群众提供了一个生动的园林科技文化互动平台。园区通过实施生态修复，变废为宝，将昔日人迹罕至的垃圾堆变为充满人气的城市公园，成为推动绿色、循环、低碳发展的成功实践。

北京园博会将城市与水岸缝合，形成紧密联系，进一步推动首都城南发展行动计划"水岸经济带"的建设发展，加快与之配套的道路、轨道交通、电力等基础设施建设，提高了北京西南部地区道路交通建设水平，改善了周边百姓的出行问题，为推动北京西南部地区的永续发展奠定了基础。同时推动永定河绿色生态走廊的建设，全面提升环境品质，改善市民休闲环境，促进城市南部地区生活质量全面升级。

SITE PLAN
总平面图

Protection and Renewal Plan of Shichahai Neighborhood in Beijing (2013-2030)

什刹海街区保护与更新发展规划（2013—2030）

Location: Beijing, China
Architect (Studio): China IPPR International Engineering Co., Ltd.
Design: 2015.10
Construction: Being Constructed
Site Area: 589ha.

地点：中国北京市
建筑师（事务所）：中国中元国际工程有限公司
设计时间：2015.10
建造时间：建造中
用地面积：589hm²

Located at the core area of old Beijing downtown, Shichahai is one of the historic districts with most humane and natural landscape. However, there are some problems, such as high population density, poor living environment, poor infrastructure and congestion, to impede the renewal and development of Shishahai.

The plan is based on the protection and inheritance of historic culture, the improvement of service facilities, the enhancement of environment and quality, the improvement of historic culture and commercial value, the promotion of organic renewal to implement sustainable development.

To relieve population problem is the primary target. The plan makes state-owned courtyards as the relocated property resources to create the mode of the integration of moving outwards and enhancing inwards. The integration of industrial upgrading and cultural protection is the main goal. The plan is to introduce the cultural industry of intangible cultural heritage to promote the development of tourism. Therefore, the historic culture would be inherited and the industry would be upgraded at the same time.

Moreover, the implementation strategies are made to realize all-win. Shichahai will be managed by governments, markets and citizens with the idea of "district operation".

什刹海街区地处北京旧城核心区，是北京城内人文与自然景观最为丰富的历史街区之一。目前什刹海街区人口密度高，居民居住环境差，市政设施条件差，交通拥堵，街区面临着更新与发展的问题。

什刹海街区保护与更新发展规划以保护和继承历史文化为前提，通过完善配套服务设施，改善街区品质与环境，提升历史文化底蕴和传统商业价值，推动街区有机更新，实现可持续发展。

1. 关注民生，实行人口有机疏解。规划利用国有院落作为回迁平移安置房源，形成"外迁疏解+平移改善"相结合的疏解模式。

2. 注重产业升级与文化保护的契合。植入非遗创意文化产业，在促进地区业态升级的同时，继承和发扬街区历史文化，带动旅游产业发展。

3. 制定实施策略，实现多方共赢。规划提出由政府、市场和市民共同管理城市公共事务的新模式，引入"街区经营"的概念。

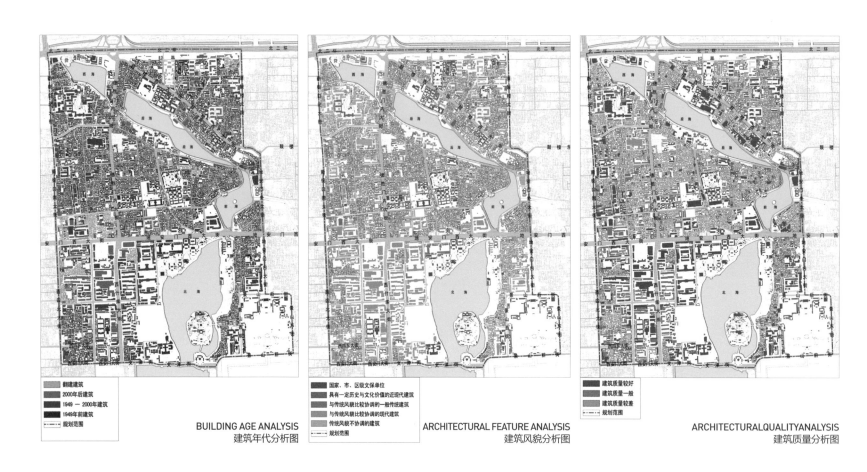

BUILDING AGE ANALYSIS
建筑年代分析图

ARCHITECTURAL FEATURE ANALYSIS
建筑风貌分析图

ARCHITECTURAL QUALITY ANALYSIS
建筑质量分析图

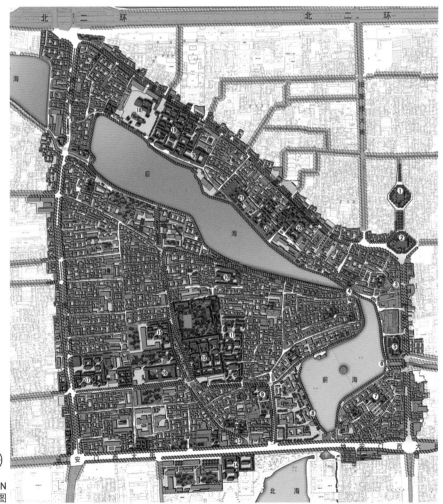

SITE PLAN
总平面图

Site Museum of Jinling Grand Bao'en Temple 金陵大报恩寺遗址博物馆

Location: Nanjing, Jiangsu Province, China
Architect (Studio): Han Dongqing, Chen Wei, Wang Jianguo, Ma Xiaodong, Meng Yuan, Southeast University School of architecture; Architects & Engineers Co., Ltd. of Southeast University
Design: 2011-2013
Construction: 2013-2015
Site Area: 75,300m²
Floor Area: 60,800m²

地点：中国江苏省南京市
建筑师（事务所）：东南大学建筑学院、东南大学建筑设计研究院有限公司；韩冬青、陈薇、王建国、马晓东、孟媛
设计时间：2011-2013
建造时间：2013-2015
用地面积：75300m²
建筑面积：60800m²

Jinling Grand Bao'en Temple was a royal temple rebuilt in Yongle, Ming Dynasty, based on the temple ruins that can backtrack to Song Dynasty. Inside the temple there was the world famous Glazed Pagoda holding Buddha relics, known as one of the seven wonders of the Middle Ages. However, the temple was ruined in the war in the mid-nineteenth century. In 2011, after years of archaeological excavations, research, competition, adjustment and demonstration by scholars from different fields, a design scheme of Jinling Grand Bao'en Temple Site Park, which is located outside of Zhonghua Gate in the south of Nanjing city, was finally formed by Wang Jianguo, Chen Wei and Han Dongqing. Jinling Grand Bao'en Temple Site Museum is the first phase project of the site park scheme, responding to two key questions. One is how to appropriately present the site information under strict protection regulations and coordinate with the multi-functional requirements of a modern museum. The other is how to reflect the relationship between history and the present using architectural form and style? The inheritance and innovation of geographical context and temporal association is achieved through the tactics of city scope ruins clustering, strata information superposed judging, site presentation oriented space management and technology innovated image reproduction. This project is innovative especially in its successful balance and interaction between ancient and contemporary, reality and illusion, demand and innovation, architecture and city.

金陵大报恩寺是明代永乐年间在原宋朝寺庙范围基础上兴建的皇家寺庙。寺庙内藏有佛祖舍利的琉璃塔曾被誉为中世纪七大奇观之一，享誉世界。该寺庙于19世纪中叶毁于战火。金陵大报恩寺遗址公园位于南京市城南古中华门外，规划设计经历众多学者长期的考古发掘、研究、竞赛、调整和论证，至2011年，在王建国、陈薇、韩冬青联合主持下基本定案。

金陵大报恩寺遗址博物馆是遗址公园的一期工程，其设计理念基于两个关键问题：其一，如何在严格的遗址保护要求下，使遗址本体的信息得到最恰当的呈现，并与现代博物馆的多元功能相得益彰。其二，如何在形式风貌上恰当地建立起历史与当下的关联。建筑创作通过置于城市格局中的遗址连缀、地层信息的叠合判断、围绕遗址展陈的空间经营和基于技术创新的意象再现等策略，实现了在地脉和时态的关联中传承和创新的初衷。新塔的创意体现于四个方面：在历史和当代之间跨越；在真实和意境之间穿梭；在需求和创新之间平衡；在建筑与城市之间互动。

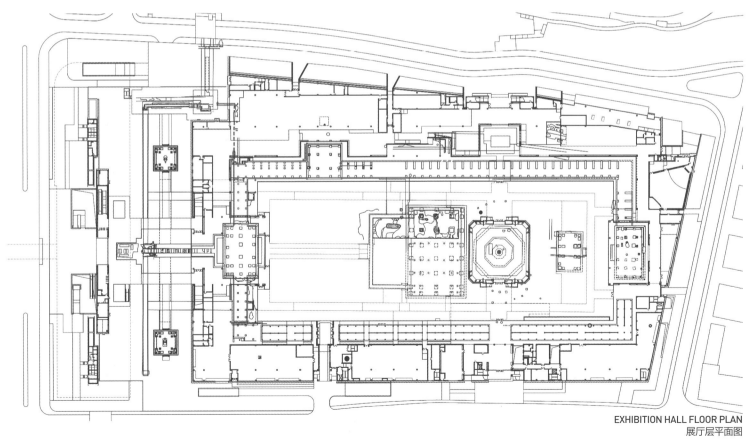

EXHIBITION HALL FLOOR PLAN
展厅层平面图

The Dinosaur Egg Remainder Museum in Qinglong Mountain

青龙山恐龙蛋遗址博物馆

Location: Shiyan, Hubei Province, China
Architect (Studio): Li Baofeng, Architecture Design institute of Huazhong University
Design: 2011
Construction: 2012
Site Area: 5,000m²
Floor Area: 1,000m²

地点：中国湖北省十堰市
建筑师（事务所）：华中科技大学建筑设计研究院：李保峰
设计时间：2011
建造时间：2012
用地面积：5000m²
建筑面积：1000m²

Being inspired by the theory of phenomenology, the designers studied special problems presented in the site, and proposed some design principles such as adapting for the local climate, highlighting historical memory, using suitable technologies, caring about presentation mode and displaying the exhibits in a relatively dark environment. Design methods: using local bamboo scaffold as concrete stencil plate, using old tiles that were taken from demolished old house nearby to cover the roofing, making double layered blinds to let wind in and block sunlight, breaking the extra-long volume into segments and fitting them to the complex topography of the site.

青龙山白垩纪恐龙蛋遗址博物馆设计受现象学理论启发，作者研究了恐龙蛋遗址所呈现的特殊问题，提出关注场所、适应地域气候、保留历史记忆、采用适宜技术、形式源于展示模式、用暗环境突出展品等设计原则，采用竹跳板作为混凝土外模板、将旧瓦作为第二层屋面、设置双层百叶实现通风阻光目标、将超长体量化整为零等设计手法，使得遗址博物馆紧密而必然地锚固于恐龙蛋遗址上。

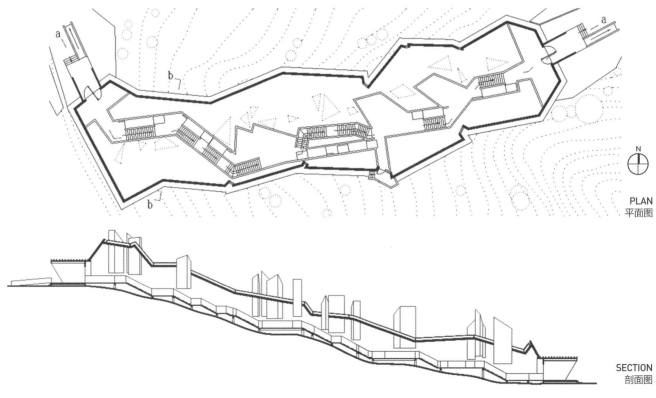

PLAN 平面图

SECTION 剖面图

The Restoration of Holy Trinity Church

基督教圣三一堂修缮工程

Location: Shanghai, China
Architect (Studio): East China Architectural Design & Research Institute, Arcplus Group PLC
Design: 2005.05-2007.04
Construction: 2007.04-2010.04
Site Area: 3,500m²
Floor Area: 2,265m²

地点：中国上海市
建筑师（事务所）：华建集团华东建筑设计研究总院
设计时间：2005.05-2007.04
建造时间：2007.04-2010.04
用地面积：3500m²
建筑面积：2265m²

Holy Trinity Church adopted the traditional Latin-cross plan layout. It is a typical British gothic revival style architecture constructed with masonry walls and timber structure. The external wall is constructed with red bricks that were measured in British measurement of inches. The porch columns are made of granite, the roof truss is made of pines and the roofing is covered with oil shade tiles. Consistent construction methods and building materials are used for the bell tower and steeple above it. "Reveal the historical truth" is the core concept of the design. Architecture restoration and authentication of historical information work are carried out simultaneously. We identify and remove all structures that latterly added on to the original building in previous refurbishments. We restore architecture and decorative ornament as accurately as possible in reference to the historical data of that time.

Similar materials and craftsman are used during restoration to keep consistence with its original style wherever possible. The goal is to ensure the restoration work to embody the historical authenticity. We keep the existing architectural parts that do not affect structure safety.

圣三一堂平面布局为传统的拉丁十字式，外观为砖墙加木屋架的英国哥特复兴式建筑。教堂墙体由英制尺寸红砖砌筑而成，侧廊柱为花岗岩石柱，屋架为洋松木，屋面外覆油页岩瓦。钟楼也是砖木结构，钟楼尖塔用与墙体同样的砖砌筑而成。本次修缮工程的核心理念是"找寻历史的真实"，对建筑本体的更新和对历史信息的鉴别成为整个工程的一体两面。一方面需要将原建筑上附着的各历史时期加建物谨慎地去除，另一方面又需要尽可能准确地复建出特定历史年代的典型装饰。修缮过程中，尽可能使用与原始建造相同或相近的建筑材料、建造工艺，对不影响结构安全的历史痕迹适当予以保留，从而尽可能使修缮的结果具有历史原真性和可辨性。

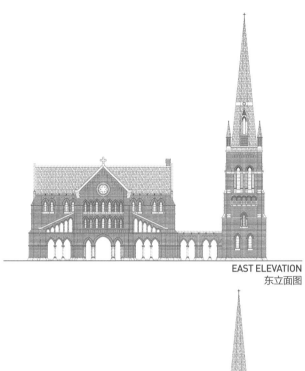

EAST ELEVATION
东立面图

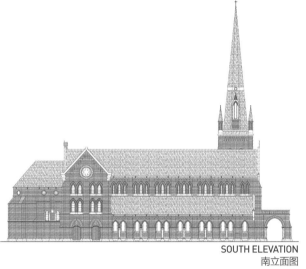

SOUTH ELEVATION
南立面图

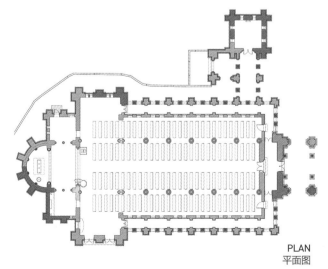

PLAN
平面图

Preservation and Restoration of the Joint Trust Warehouse

四行仓库修缮工程

Location: Shanghai, China
Architect (Studio): Institute of Shanghai Architectural Design & Research (Co., Ltd.), Arcplus Group PLC
Design: 2014.07-2015.08
Construction: 2015
Floor Area: 25,550m²

地点：中国上海市
建筑师（事务所）：华建集团上海建筑设计研究院
设计时间：2014.07-2015.08
建造时间：2015
建筑面积：25550m²

The project consists of two joint warehouses on the shore of Suzhou Creek. The one on the west is Joint Trust Warehouse, which was built in 1935 and was designed by Atkinson & Dallas Ltd. The main structure of this five-story building is a reinforced concrete slab-column structure. In the Battle of Shanghai in 1937, the well-known "Lone Battalion" took it as the final defense stand against the Japanese troops before retreat. Now the warehouse is a designated historic site on the municipal level to memorize this battle.

History is highly respected and authenticity is taken as design principle in preservation and restoration work. In the battle of 1937 the western wall of the warehouse was torn by the explosive shells. Several methods have been taken to locate the exact positions of the damaged parts and make sure they are revealed in a way as same as in the battle of 1937. At the same time various new techniques have been used to ensure safety of the building. The 7th floor – an added part to the original building – is removed. The external wall of 6th floor is made to recess so that the north and south facades could retain the original character. The central passage is restored and used as a characteristic atrium. Part of the west side of the atrium is used as a "Memorial Hall for Anti-Japanese War" to demonstrate the significance of this historic site, while the quality of the rest is highly improved to meet the demand of "office for creative industries".

苏州河畔的四行仓库由两座仓库组成，西部的"四行仓库"建于1935年，由"通和洋行"设计，原高五层，主要为钢筋混凝土无梁楼盖结构体系。这里是1937年抗日战争中著名的"四行仓库保卫战"的发生地，如今是上海市市级文物保护单位。

本次保护与复原设计以尊重历史、真实性为原则，用多种方法查明西墙在抗战时的炮弹洞口位置，力求准确复原梁柱边的洞口并采取多种创新技术确保建筑安全；拆除后期搭建的七层，使六层后退，恢复南北立面历史原貌；恢复原中央通廊特色空间，改作中庭，其西侧设立"抗战纪念馆"，彰显抗战遗址历史意义；其余部分提高舒适度，用作创意办公等空间。

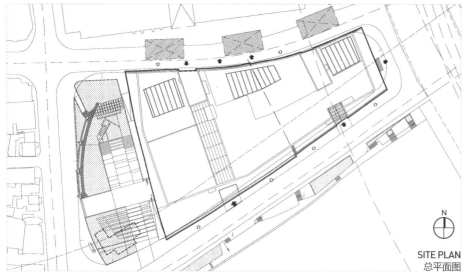

SITE PLAN
总平面图

WEST ELEVATION
西立面图

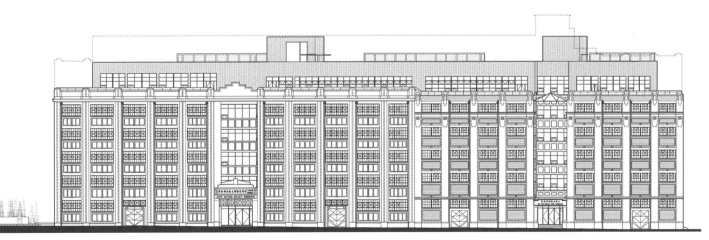

SOUTH ELEVATION
南立面图

Bund 33 # Renovation of the Original British Consulate Building and the Apartment

上海外滩源 33#
原英国领事馆及官邸历史建筑保护及再利用工程

Location: Shanghai, China
Architect (Studio): East China Urban Architectural Design & Research Institute, Arcplus Group PLC
Design: 2007-2010
Construction: 2009-2012
Site Area: 22,000m²
Floor Area: 20,000m²

地点：中国上海市
建筑师（事务所）：华建集团华东都市建筑设计研究总院
设计时间：2007-2010
建造时间：2009-2012
用地面积：22000m²
建筑面积：20000m²

Bund 33 # is at the north of Suzhou River, covering an area of about 22,000m². The project consists of four parts: the public green space of the Bund and the underground space, the renovation of the original British Consulate Building and the apartment, the restoration of the Union Church and the church apartment, the renovation of the Rowing Club history building.

The design concept of Bund 33 # project is "To restore the style, to remodel function".

1. Protect and retain the historical buildings, restore the historical style of the street space pattern, and retain the original British Consulate main building and the official residence of brick include wood structure; restore the official residence of the open porch, Corinth column and column colonnade and exquisite decoration;

2. Enhance the landscape area to be more open and public, and connect it to the city green space;

3. Use scientific and standardized construction method, protect the ancient trees for security.

"To remodel function" includes:

1. Improve the regional functions, include cultural, exhibitions, financial Expo, performing arts, leisure and artistic experience of the place;

2. Strengthen the publicity and openness of the regional greening, so as to connect with Suzhou riverside space;

3. Construct underground space, to ensure that substations, overhead lines and other facilities placed underground, and build underground garage to settle the parking problem in the region.

外滩源33#北临苏州河，占地面积约2.2万m²。项目由外滩源公共绿地及地下空间利用、原英国领事馆主楼和官邸修缮工程、原联合教堂和教会公寓修缮工程以及原划船俱乐部历史建筑修缮工程四个部分组成。

外滩源33#项目以重现风貌、重塑功能为修缮设计理念。

重现风貌包括：

1.保护和保留历史建筑，恢复街道空间格局的历史风貌，保留原英领馆主楼及官邸砖木结构；还原官邸连续的敞廊、科林斯柱式和组合柱式的柱廊及精美的装饰；

2.还原项目区域景观风貌，还原城市绿化空间；

3.科学规范施工，保护百年古树名木的安全。

重塑功能包括：

1.完善区域功能，形成文化展示、金融博览、演艺、休闲和艺术体验的场所；

2.强化区域绿化的公共性和开放性，同苏州河滨水绿地连接；

3.地下公共空间的建设，确保变电站、架空电线等设施置于地下，并兴建地下车库缓解该区域停车问题。

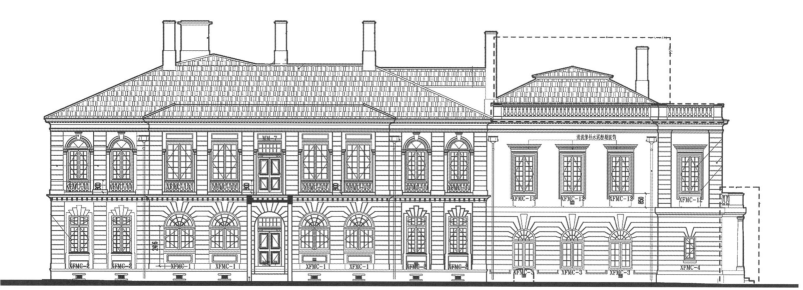

NORTH ELEVATION
北立面图

Protection Plan of Fujian Tulou, the World Cultural Heritage

世界文化遗产 福建土楼 保护规划总纲

Location: Nanjing County, Zhangzhou, Fujian Province, China
Architect (Studio): Architectural Design and Research Institute of Tsinghua University Co., Ltd.
Beijing Tsinghua Urban Palnning& Design Institute
Design: 2010.09-2013.02
Site Area: 10,870,000m²

地点：中国福建省漳州市南靖县
建筑师（事务所）：清华大学建筑设计研究院有限公司、北京清华同衡规划设计研究院有限公司
设计时间：2010.09-2013.02
用地面积：10.87km²

The "Protection Plan Fujian Tulou, for World Cultural Heritage " is an outline document for protecting the heritage of Fujian Tulou (earth towers). It further defines the heritage composition and highlights its universal value. Based on the objective analysis of the reality of the heritage site, the protection plan functions as a summed up and improved culmination of all the existing protection planning in a holistic manner. It meets the requirements related to the protection of world cultural heritage, and also offers solutions for major problems regarding the protection, management, display, utilization and research of the Tulou. It provides guidelines and macro-strategies for protecting 10 heritages sites in the 12th five year plan period and the longer period beyond that.
The planning is novel and distinctive in terms of:
1. It is the first time that the concept of sustainable development is introduced in the protection plan for a world cultural heritage. Building on the general sense of sustainable development in economic, environmental and social realms, it innovatively puts forward the idea of a culturally sustainable development on the basis of the latest theories in the international heritage protection community.
2. The traditional community at the heritage site is a breakthrough concept in this planning, leading to an in-depth evaluation and analysis on the connotation of traditional community, the impact of being a world cultural heritage on such communities and the significance of traditional communities to heritage protection.

《世界文化遗产福建土楼保护规划总纲》作为"福建土楼"遗产地保护的纲要性文件，进一步明确了福建土楼的遗产构成及突出普遍价值，并在对遗产地现实情况进行客观分析的基础上，从整体着眼，对已有的各类保护规划进行了归纳统筹和总结提升，使其符合世界文化遗产保护的相关要求，并对"福建土楼"遗产保护、管理、展示利用及研究等领域面临的主要问题提出了解决方案，为10处遗产地在"十二五"及更长的时期内保护管理工作的开展提供了原则性指导和宏观策略。
规划的创新与特点主要体现在以下两个方面：
1. 在世界文化遗产保护规划中首次系统引入了可持续发展的概念，并在一般意义上的经济、环境和社会的可持续发展之外根据国际遗产保护界的最新理论成果提出了文化的可持续发展。
2. 规划提出了遗产地传统社区的概念，并围绕传统社区的内涵、传统社区在申遗成功后受到的影响以及传统社区对遗产保护的意义做了详细的评估与分析。

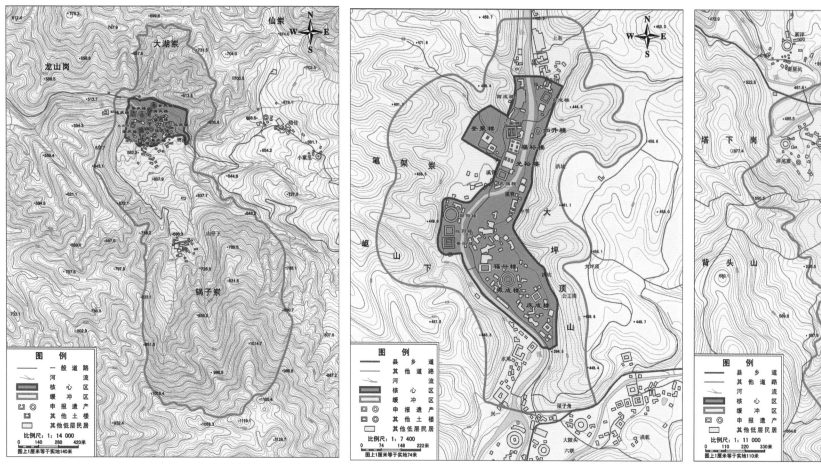

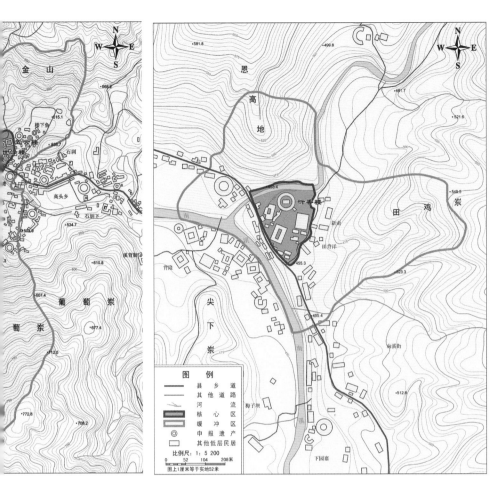
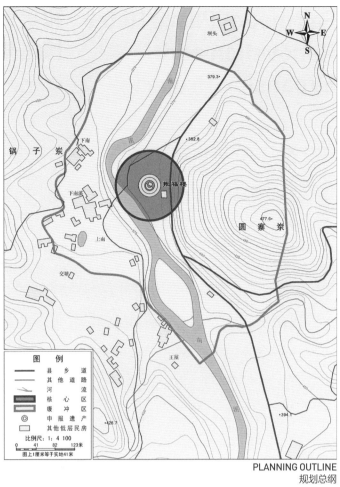

PLANNING OUTLINE
规划总纲

Fairmount Peace Hotel Renovation and Expansion Project

和平饭店修缮与整治工程

Location: Shanghai, China
Architect (Studio): Institute Of Shanghai Architectural Design & Research (Co., Ltd.), Arcplus Group PLC
Design: 2007-2010
Construction: 2009-2010
Floor Area: 51,149m²

地点：中国上海市
建筑师（事务所）：华建集团上海建筑设计研究院
设计时间：2007-2010
建造时间：2009-2010
建筑面积：51149m²

Northern Building of Peace Hotel, formerly known as "Sassoon House", was built by Victor Sassoon and designed by Palmer & Turner Architects. Completed in 1929, the building was appraised as "The First Building in the Far East" back then, and is now one of the most famous modern architecture masterpieces in Shanghai. The building is among the high-rise buildings in China with a height of about 70 meters, marked by iconic parts of decorative artistic style and Art Deco interior, with the coexistence of different styles.

The most comprehensive conservation and renovation of this Building in history was completed during 2007-2010. In order to improve the configuration standards and functions of hotel, new buildings were constructed at the west side of the inner court.

The design features are as follows:
1. The original appearance and function of important indoor and outdoor spaces of the historic building was protected and restored, so as to preserve its completeness and authenticity to the best possible extent. For example, the "丰-shaped" corridor and "octagonal atrium" on the ground floor.
2. Key historical parts are taken good care of.
3. The overall structure of the building was fortified, with improved performances in fire protection, energy-efficiency, environmental protection, health and epidemic prevention.
4. New structures were expanded to improve high-end hotel functions and house new facilities.
5. Comfort level of guest rooms was raised with more advanced functions.

和平饭店北楼原名"沙逊大厦"，由维克多·沙逊筹建，公和洋行设计，1929年竣工，曾有"远东第一楼"的美誉，是上海最著名的近代建筑之一。

大楼为高近70m的早期高层建筑，重点部位呈装饰艺术风格，内部装饰以Art Deco为主，多种风格并存。

2007-2010年对和平饭店北楼进行历史上最全面的保护修缮。为提高酒店配置标准与功能，在原西侧内院扩建新楼。

设计特点为：
1. 保护恢复老楼重要室内外空间历史原貌与原有功能。最大限度保存其完整性、真实性，如：底层"丰"字形廊、"八角中庭"等。
2. 精心保护各重点部位。
3. 结构加固，提升大楼消防、节能、环保、卫生防疫性能。
4. 扩建新楼，完善高端酒店功能，新增设施。
5. 提高客房舒适度，功能现代化。

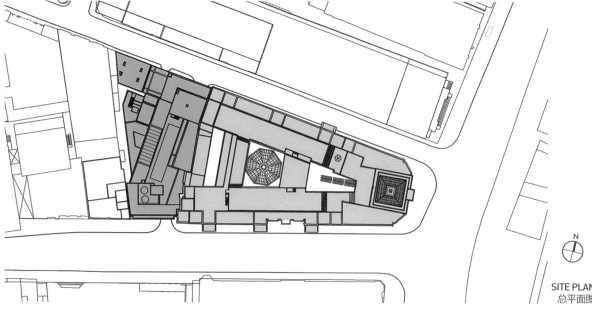

SITE PLAN
总平面图

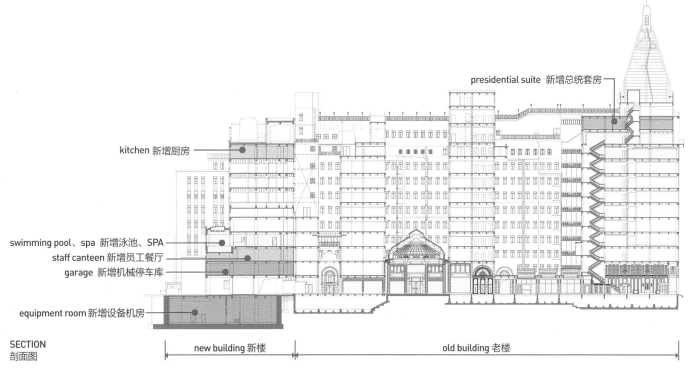

SECTION
剖面图

presidential suite 新增总统套房
kitchen 新增厨房
swimming pool、spa 新增泳池、SPA
staff canteen 新增员工餐厅
garage 新增机械停车库
equipment room 新增设备机房

new building 新楼 | old building 老楼

Preservation and Reparation Project of Shanghai Great World

上海大世界修缮工程

Location: Shanghai, China
Architect (Studio): East China Urban Architectural Design & Research Institute, Arcplus Group PLC
Design: 2004.10-2017.03
Construction: 2009.10-2017.03
Site Area: 31,893m²
Floor Area: 16,626m²

地点：中国上海市
建筑师（事务所）：华建集团华东都市建筑设计研究总院
设计时间：2004.10-2017.03
建造时间：2009.10-2017.03
用地面积：31893m²
建筑面积：16626m²

Ever since its completion in 1917, Shanghai Great World is located at the most prosperous area in downtown of Shanghai, with the honour of the Biggest Amusement Park in the Fareast.

In year of 1989, the Great World was announced as Excellent Modern Preservation Building in Shanghai, and underwent repairment from 2003 to 2017.

The historical façade of the building during year 1931 to year 1952 was chosen as the blueprint of the repair work. The original part was preserved as much as possible, and the added part was easily recognized from the old ones.

As for the inner space and function renewal, lots of activities that were welcomed by the public were placed inside the building to make sure that all the memories and feelings of citizens can be kept.

After renovation, Shanghai Great World was open to the public, being regarded as a non-material cultural heritage exhibition center.

自1917年诞生以来，上海大世界就坐落于上海最繁华的商业地段，有着"远东第一大游乐场"的美誉。

1989年，大世界主体建筑被上海市人民政府公布为"上海市优秀近代保护建筑"。2003年起停业整治，2017年修缮完工，重新对外开放。

设计选取黄金荣时期（1931—1952）的大世界外立面作为改造修复的蓝本，最大限度保留建筑原存部位，修复与新增的部分也与旧有部分有所识别，避免以假乱真。

对于建筑室内格局及其功能，设计师置入公众喜爱的各类活动来带动大世界的改造更新，而通过尊重原真性的建筑本体作为烘托活动的场所，市民的记忆与情感始终参与在大世界的改造项目中。最终，大世界以非文化遗产展示中心的性质重新对外开放。

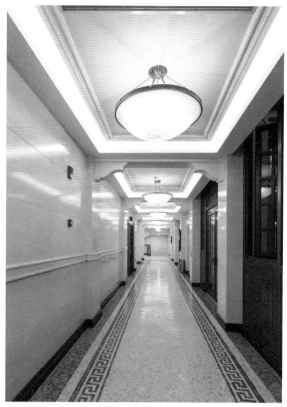
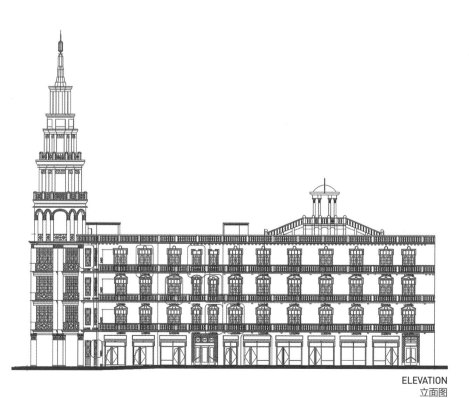

ELEVATION
立面图

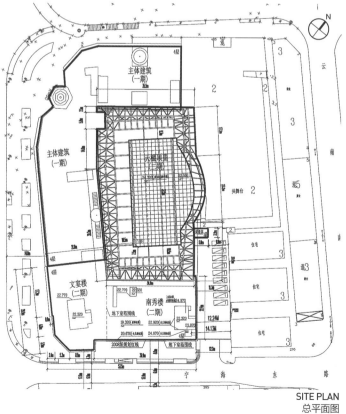

SITE PLAN
总平面图

Preservation Project of Prince Gong Mansion
恭王府府邸文物保护修缮工程

Location: Beijing, China
Architect (Studio): Architectural Design and Research Institute of Tsinghua University Co.,Ltd.
Design: 2004.10-2008.08
Construction: 2008.12
Site Area: 32,000m²
Floor Area: 12,600m²

地点：中国北京市
建筑师（事务所）：清华大学建筑设计研究院有限公司
设计时间：2004.10-2008.08
建造时间：2008.12
用地面积：32000m²
建筑面积：12600m²

Prince Gong Mansion is a State Protected Historic Site located at Liuyin Street in the Shichahai area of Beijing. It is one of the best preserved ancient royal mansions. The Prince Gong's Mansion has sophisticated arrangement and magnificent architecture, and is no doubt one masterpiece among all princes' mansions (Wangfu) in the Qing Dynasty. It bears rich historical and cultural connotation as well as high architectural and aesthetic value, and is said to be "a mansion that tells half of the Qing Dynasty history". As time passed by, the Mansion had been severely damaged and dilapidated. During this project, a major maintenance will be carried out, which aims at restoring the old glory of the mansion and revitalizing it into the first national Wangfu museum that fully displays all sides of Qing Dynasty's Wangfu culture including related standards and patterns, architecture, ornaments, artifacts, etc.

This renovation project is the most scientific, thorough and elaborate renovation project since the foundation of the PRC. The project design is carried out strictly according to both domestic and international theories and principles on historic preservation, such as the "Law of the People's Republic of China on the Protection of Cultural Relics", "Protection Regulations for Chinese Cultural Relics Sites", and the "Venice Charter". With a consistent commitment to a design method that combines scrupulous academic research and design & construction, the project is carried out with a strong focus on details and refinement in order to render a more accurate and scientific restoration design.

恭王府为国家级文物保护单位，坐落于北京什刹海的柳荫街，是现存较为完整的王府古建筑群。曾经的恭王府府邸格局精严，建筑华美，堪称清代诸王府第的杰作。所谓"一座恭王府，半部清朝史"，它承载了丰厚的历史文化内涵和珍贵的建筑美学价值。随着时代的更迭变迁，府邸早已损毁严重、破败不堪。本工程的目标就是要通过这次百年大修，重现恭王府昔日的辉煌，同时将恭王府建成第一座国家级的王府博物馆，充分展示清代王府的规制、建筑、装饰和器物以及王府文化的诸多方面。恭王府府邸文物保护修缮复建工程是新中国成立后对恭王府府邸最科学、最彻底、精度最高的一次文物修缮。工程设计严格按照《中华人民共和国文物保护法》和《文物建筑保护准则》以及《威尼斯宪章》等国内、国际的文物保护法规、理论和原则进行。本工程始终坚持严谨细致的学术科学研究与设计、施工相结合的设计方法，关注细节、精益求精，使复原设计更准确、更具科学性。

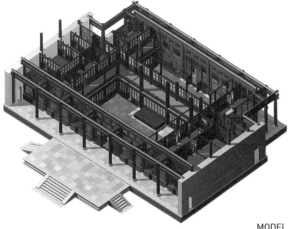

MODEL
模型图

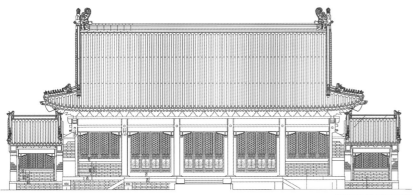

SOUTH ELEVATION
南立面图

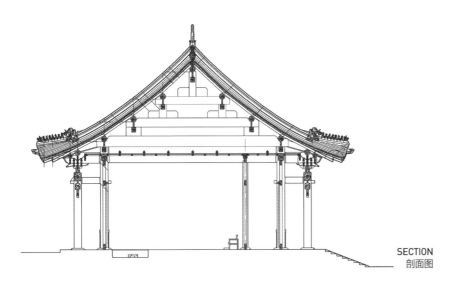

SECTION
剖面图

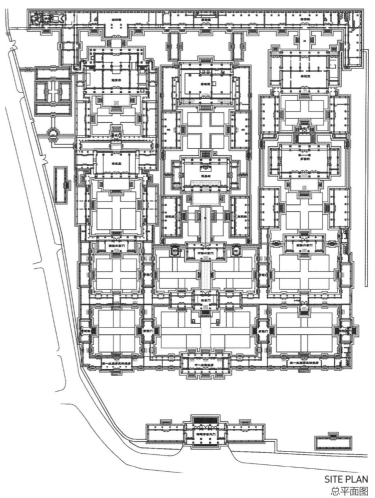

SITE PLAN
总平面图

Restoration of the Building of Shanghai Kunju Opera Troupe

上海昆剧团大楼修缮改造工程

Location: Shanghai, China
Architect (Studio): Zhang Jiezheng, Chen Minsheng, Zhen Ning, Fu Yong (Historic Building Conservation Design & Research Institute, Arcplus Group PLC)
Design: 2009-2010
Construction: 2011-2013
Site Area: 2,335m²
Floor Area: 3,293m²

地点：中国上海市
建筑师（事务所）：华建集团历史建筑保护设计院：张皆正、陈民生、郑宁、付涌
设计时间：2009-2010
建造时间：2011-2013
用地面积：2335m²
建筑面积：3293m²

Located at No.9 Shaoxing Road, the building of Shanghai Kunju Opera Troupe was designed in 1932 and completed in 1935. The building was originally used as a social club for the French army and policemen, and was later converted into a policeman's museum, documenting the history and development of modern policeman as a profession. Since the founding of the P.R.C., the building has been used as the rehearsing and office space for recreational troupes. It has witnessed the development of the various opera troupes, bearing critical social and cultural values. In 2005, the building was listed as a fourth-batch Shanghai Historic Buildings, and falls into second protection category.

This is an early modern-era building with partial Art-Deco features. The facade is covered in beige plaster with Art-Deco architrave details. Its roof with the gentle slope was thoughtfully integrated with the terrace.

The interior spatial layout of the building primarily serves to meet the functional needs, with a classic and gorgeous foyer. The spaces are generously lit by natural light. The most characteristic of all is the multi-colored terrazzo floor, the delicate crown moldings, and the light fixtures in Art-Deco style.

The restoration of the building facade reflects the historical value of the building. With the aim to preserve in mind, the interior spatial layout has been optimized, and the spatial sequence has been reorganized. The restoration project is a delicate balance between the functional needs of the Kunju Opera Troupe to rehearse and to promote its intangible cultural heritage, and physical improvement of the building's envelope, performance, and its MEP systems. Especially worthy of note is the careful restoration of the primary spaces such as the foyer, hall, rehearsing hall, staircase, etc.

The design highlights inheritance of the architectural heritage, and the feature of original spatial layout and decorative details.

绍兴路9号上海昆剧团大楼于1932年设计，1935年建成。初为法国军人之家和警察俱乐部；后改做警察博物馆，记录了近代警察职业发展历史；新中国成立以来，大楼作为文艺团体的排演办公场所，见证着海上梨园艺术的传承发展，具有重要的社会文化价值。2005年被公布为上海市第四批优秀历史建筑，二类保护。

这是一幢局部带有装饰艺术风格特征（ART-DECO）的早期现代建筑。立面采用浅米色抹灰饰面，缓坡顶与露台相结合，局部点缀有装饰艺术风格线脚。室内布局以功能为先、空间摩登、采光充分，门厅则古典气派。最具特色的是彩色水磨石楼地坪和同样具有装饰艺术特征的线脚以及灯具。

外立面风貌的保护注重体现年代价值；空间优化与流线梳理以保护为先，兼顾昆曲艺术非物质文化遗产的高雅艺术文化传播与排演功能需求；全面提升建筑品质、性能，更新设备设施。特别是对门厅、前厅、排练厅、楼梯间等主要空间的修缮复原，注重对建筑文化遗产的格局装饰特色价值的挖掘与传承。

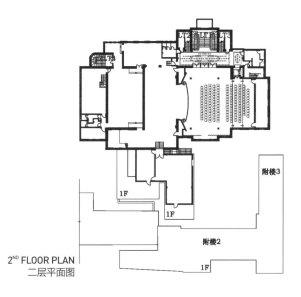

2ND FLOOR PLAN
二层平面图

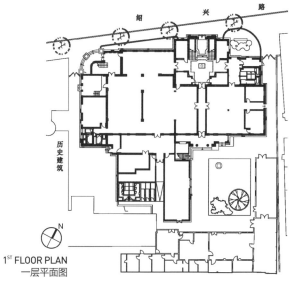

1ST FLOOR PLAN
一层平面图

Rural Reconstruction

乡村建设

China since ancient time has been taking agriculture as country's base. The integration of cities and the villages in ancient society can be interpreted by the idea of the same structure of clan and country. Reliable social and economic bonding linked cities and villages, which formed a unique and sustainable living environment with high efficiency and great diversities.

In the process of modernization, the rural economy and culture values that have been nourishing Chinese people over thousands of years are deteriorating, and natural villages are perishing or have disappeared. The social foundation that once bonded villages and cities on longer exists. More differences other than isomorphism can be found between them in many aspects.

However, when urban life style prevails and is getting standardized, people today more than ever aspire to have previous pastoral life in villages. Rural construction has become a livelihood topic that attracts extensive attention in China. Following the country's New Urbanization strategy, we highlight small village development and ancient village protection; we ensure an overall rural development and the involvement of local residents; we partially upgrade the rural construction according to local conditions without changing the original features; we unify rural economic, cultural and ecological values through environmental reform, industries support and social organization restructuring, so as to achieve the sustainable development goal.

Protection and Regeneration of Traditional Villages
There are tangible and intangible cultural heritages in nationally protected traditional villages. They are carriers of traditional Chinese cultural essence, rich in historic and cultural implications. They are legacies of Chinese farming civilization, and have high values in history, culture, science arts, society and economy. Our goal is to protect and revive them by retaining their ancient spatial textures and architectural style, and revitalizing their spatial programs.

Construction and Renewal of Beautiful Villages
China is vigorously promoting Beautiful Village Construction that highlights ordinary villages and towns as it covers much more extensive areas than the listed national heritage villages. With goals to retain unique features of villages and restore their self-generation functions, the Construction emphasizes a dynamic balance in sustainable development for production, living and ecology through master-planning and upgrading the villages' spatial patterns and textures, architectural styles, infrastructures and environment.

Rural Architecture
In addition to master-planning and upgrading villages, another major aspect is to newly build or partially rebuild the constructions, a micro intervention like acupuncture treatment. It can be new cultural facilities like farmer's assembly hall and library, or be tourism facilities like newly built bed and breakfast and museum that take advantage of natural and cultural resources. They facilitate fusion of city and village, and contribute to a mutually rewarding circulation.

中国自古以农立国，传统社会中的城乡一体就是家国同构、天下归一思想的物质体现，它以城乡之间稳定可靠的社会经济纽带为基础，以基于血缘—地缘关系的双重管理制度为保障，形成了独特、持续而高效的多样性生活环境。千百年来，无数脍炙人口的诗句描绘出城乡一体的优美图景。

在近现代化的过程中，曾经滋养了中国数千年人文历史的乡村经济、文化价值逐渐弱化，自然村落正在消失或消亡，古代城乡一体的社会基础不复存在，城乡之间在各方面的差异性都大大超过了同构性。

然而，在城市生活标准化、普及化后的今天，人们对留住乡愁、回归田园的渴求愈加强烈。乡村建设成为中国广泛关注的民生主题，重建城乡一体的文化认同，梳理城乡之间的文化脉络，将已经分离的城与乡再度弥合起来，是城乡发展与建筑创新的重要方向。在国家"新型城镇化"战略背景下，小村镇建设和传统村落保护成为乡村建设的重点，注重乡村地区的整体发展和乡村居民的积极参与，以保持原有的乡村特色为主，因地制宜地进行整体保护梳理和局部的建设更新，通过环境改造、产业帮扶和社会组织重构以实现乡村经济价值、文化价值和生态价值的统一，达到乡村可持续发展的目的。

传统村落的保护与再生

国家重点保护的传统村落拥有物质形态和非物质形态文化遗产，承载着中华传统文化的精华，蕴藏着丰富的历史信息和文化景观，是中国农耕文明留下的宝贵遗产。对于这种具有较高的历史、文化、科学、艺术、社会、经济价值的村落，目标偏重于保护与再生，即保留传统空间肌理与建筑风貌，保护和振兴传统村落，实现乡村遗产空间的功能再造，使传统遗产资源得以延续和再生，实现乡村文化复兴和传承。

美丽村镇建设与更新

除国家重点保护的传统村落以外，更为量大面广的是普通村镇和村落，中国大力推进的美丽乡村建设注重生产、生活、生态的和谐发展，以保持乡村特色和实现村庄自我更新为目标，对乡村形态、空间肌理、建筑风貌、基础设施和环境等进行合理规划、更新改造和建设，以产业振兴实现可持续发展，达成乡村文化、产业、生态环境的动态平衡。

田园建筑

除了对乡村进行整体的规划和更新外，局部的新建改造或针灸式的微介入也是乡村建设的重要内容，例如修建村民会馆、图书馆等农村文化设施，提升基础设施水平，加深农民与市民之间的文化融合；或依托乡村独具特色的自然及人文景观资源，新建民宿、博物馆等文化旅游设施，将乡村旅游和环境保护相结合，形成良性发展格局。

Homeland of Mosuo, the Project of Protecting Mosuo Habitation

摩梭家园
——摩梭人聚居地保护

Location: Lugu Lake, Sichuan Province, China
Architect (Studio): China Southwest Architecture Design & Research Institute Co., Ltd.
Design: 2014
Construction: 2015

地点：中国四川省泸沽湖
建筑师（事务所）：中国建筑西南设计研究院有限公司
设计时间：2014
建造时间：2015

Mosuo Culture, a matriarchal clan culture, which is owned by the Mosuo residents who live around Lugu Lake on the border between Yunnan and Sichuan Provinces. With the development of globalization, Mosuo Culture is facing the threat of gradual demise. Protecting and reconstructing the homeland of Mosuo Culture is the first priority of this project.

The design is based on the unique maternal clan culture and their traditional modes of production. It respects Mosuo traditional modes of production and lifestyle. This project is aimed to improve their standard of living by improving the infrastructure and public environment of this villages.

The design uses a modular strategy to deal with the diversity of traditional living spaces. This project makes the four parts of traditional Mosuo residential (Grandmother's Room, Thatched Cottage, Buddha Hall and Flower House) to be precise designs for providing different standard modules. Mosuo residents can be free to combine these modules as a variety of natural growth community.

The design emphasizes the native building materials and the original construction. The project retains a large number of Mosuo original construction methods during the design and construction process, in order to reducing the traces of architects' design.

Taking the special conditions of remote areas into account, the project uses special construction modes, which are "unified planning", "native villagers constructing" and "architects guiding on the site". Mosuo residents are fully taking part into the construction and design process, to achieve the goal of inheriting and innovating Mosuo culture.

摩梭文化，是世代生活在泸沽湖畔的摩梭人所独有的母系氏族文化。随着全球化，摩梭传统文化面临逐渐消亡的威胁。保护和建设摩梭文化所依托的"家园"，是本项目的首要目的。

设计基于特有的母系氏族文化和传统生产方式的全域保护更新原则，充分尊重摩梭人传统生产生活方式，同时改善村落的公共环境，完善基础设施，提升生活水平，实现传统文化与村落共存。

应对多样性传统居住空间，设计采用了模块化策略。将传统摩梭民居"祖母屋、草楼、经堂、花楼"四部分细化设计，提供不同的模块标准，居民可自由组合，形成多样性自然生长的建筑群。

设计强调建筑材料的乡土性和建造的原真性，在设计和建造的过程中，保留大量摩梭人原始的建造手段和方式，整旧如旧，减少建筑师创作的痕迹。

针对偏远地区的特殊条件，项目采取了"统一规划设计+村民参与自建+建筑师现场控制指导"的建设模式，摩梭人充分参与其中，实现文化的传承和创新。

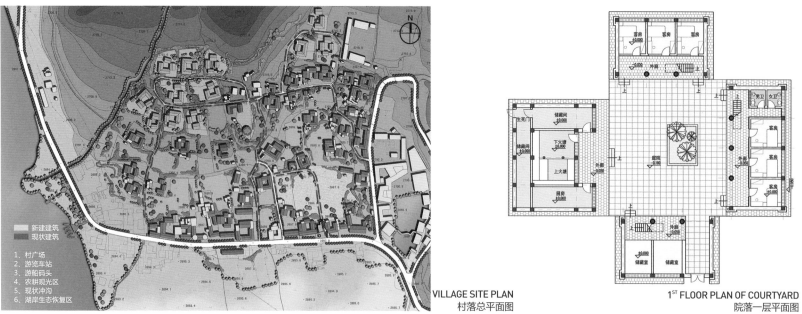

1、村广场
2、游览车站
3、游船码头
4、农耕观光区
5、现状冲沟
6、湖岸生态恢复区

新建建筑
现状建筑

VILLAGE SITE PLAN
村落总平面图

1ST FLOOR PLAN OF COURTYARD
院落一层平面图

Shaxi Rehabilitation Project

沙溪复兴工程

Location: Shaxi Town, Jianchuan County, Yunnan Province, China
Architect (Studio): Swiss Federal Institute of Technology (ETHZ)
Design: 2004
Construction: 2004-2010
Site Area: 8,150m²
Floor Area: 3,200m²

地点：中国云南省大理州剑川县沙溪镇
建筑师（事务所）：瑞士联邦理工大学
设计时间：2004
建造时间：2004-2010
用地面积：8150m²
建筑面积：3200m²

Shaxi Rehabilitation Project (SRP) takes a pioneer approach in heritage conservation and endogenous economic growth based on the notion of culture as a catalyst for development and structured to maximize on the abundant local resources. The overall ambition of SRP is the sustainable development of region based on carefully planned measures for responsible growth, best use practices maximizing on local natural, cultural, environmental and human resources, and a restitution of the past for future generations to explore centuries of cultural tradition and regional history.

SRP strategy approaches conservation and sustainable development in six different modules, which are integrated over a three-dimensional scale. The minor scale is about key heritage building conservation, ensuring that the value of historic landmarks is well preserved. The middle scale, which was considered as the background of the minor scale, applies to a sophisticated built heritage context, integrating modern infrastructure into a historical environment. And the great scale addresses a comprehensive sustainable development strategy for the whole Shaxi Valley.

Modules 1-6 based upon the three-dimensional scale, respectively, are: Marketplace Restoration, Historic Village Conservation, Sustainable Valley Development, Ecological Sanitation, Poverty Alleviation and Dissemination.

沙溪复兴工程是一个综合性的文化保护与发展项目。它以中国边远的西南山区中深厚而丰富的历史文化遗产为根基，以古建筑保护为切入点，以旅游发展为经济动力，以期实现社会、文化、经济、资源、景观之间相互依托、彼此协调的可持续发展。

沙溪复兴工程被规划为三个层次。沙溪复兴工程的第一个层次，也是最基本的层次，即核心建筑遗产层次，即如何让这些优秀的建筑遗产永久保存。这些古建筑都是漫漫历史中形成的精华，代表着各个历史阶段的文化成就；第二个层次是古村落层次，这是古建筑所依托的环境，也是民居群体价值的体现。村落的生活条件和吸引力决定了村落长久的活力。第三个层次是整个坝子，这是古村落的背景，优美的自然景观衬托出古村落的古朴，但是长远的发展必须与因地制宜的经济发展战略相结合。

以这三个层次为基础，沙溪复兴工程进一步具体化为六个项目组成部分：四方街修复、历史村落保护与发展、沙溪坝可持续发展、生态卫生、脱贫与地方文化保护、对外交流。

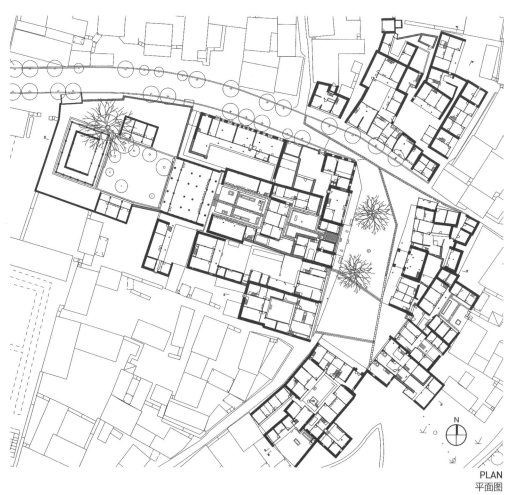

PLAN
平面图

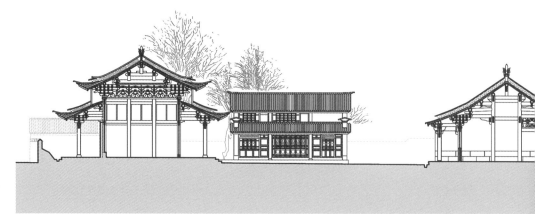

SECTION
剖面图

The Practice of Protection and Utilization of Traditional Villages in Yangjiatang Village, Songyang County

松阳县杨家堂村传统村落保护与发展实践

Location: Songyang County, Zhejiang Province, China
Architect (Studio): Zhejiang Ancient Architecture Design and Research Institute
Design: 2013
Construction: 2014

地点：中国浙江省松阳县
建筑师（事务所）：浙江省古建筑设计研究院
设计时间：2013
建造时间：2014

Yangjiatang village is located in Songyang County, Zhejiang Province. It faces west and spreads along the mountain. The overall structure and architectural style are intact. The village has a long history of more than 350 years. The protection and development of Yangjiatang village is people-oriented and follows flexible protection principles.

The priority is to protect the village based on the harmony between man and nature. The emphases are put on the overall restoration of the integration of several ecological landscape resources including mountains, water, and villages, and the optimization of the ecological environment of villages.

Improve people's livelihood, and create a comfortable living environment. To meet villagers' living needs, the living conditions of the traditional buildings should be improved, and the public services and municipal infrastructure should be improved by renovating the village environment, intercepting the sewage pipes and classifying the garbage.

Explore and inherit the characteristics of traditional culture heritage of Yangjiatang. Culture construction like the temple, the ancestral hall, school can be changed into folk museums and folk festival activity venues where folk customs can be exhibited and experienced. Refine the Yangjiatang's peculiar traditions like wall literature, family motto so that those excellent culture genes can be passed on.

Revive the economy and strengthen the inner cohesion of the village. Set up village history exhibition hall in combination with the village's administrative office. Guide the establishment of home care center and the village library. Cultivate commercial activities such as entertainment and health, characteristic B&B industry, ecological agriculture, agricultural products sales and so on, to help resurrection of the village economy.

杨家堂村地处浙江松阳，村落坐东朝西，依山就势，整体格局及建筑风貌完好，建村距今已有350余年历史。杨家堂的保护发展遵循以人为本、活态保护的原则。

优先保护"天人合一"的村落形态：重点对山、水、村落几者相融的生态景观资源进行整体修复，优化村落生态环境。

改善民生，营造宜居的生活空间：结合村民生活需求对传统建筑居住条件进行提升，通过整治村落环境、截污纳管、垃圾分类等措施完善公共服务和市政基础设施。

发掘特色，传承杨家堂的传统文化：利用社庙、祠堂、学堂等文化建筑，建设成具有展陈体验功能的民间博物馆和民俗节会活动场所，提炼杨家堂特有的传统墙头文学、居家格言，传承村落优良的文化基因。

复活经济，增强村落内在凝聚力：结合村集体办公场所设立村史展示馆，引导建立居家养老中心、村民图书室，培育休闲养生、特色民宿、生态农业、农产品售卖等业态，复活村落经济。

SITE PLAN
总平面图

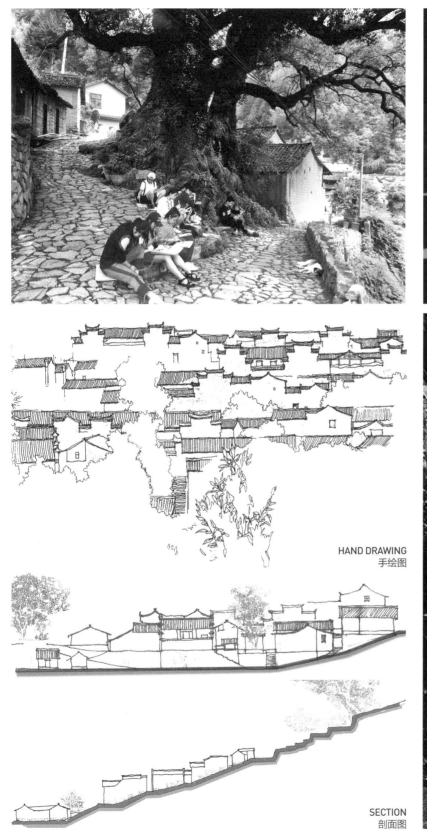

HAND DRAWING
手绘图

SECTION
剖面图

Planning of Xiedian Traditional Village Protection and Regeneration

谢店村传统村落保护与再生规划设计

Location: Xiedian Village, Macheng, Hubei Province, China
Architect (Studio): Xiao Wei, He Xinran, Yang Qingsong, Li Weiqiang, Li Jingjing, Wu Hang, Zhang Si, CITIC General Institute of Architectural Design and Research Co, Ltd
Design: 2015.07-2016.01
Construction: 2016.01-2016.09
Site Area: 110 mu
Floor Area: 9,400m²

地点：中国湖北省麻城市谢店村
建筑师（事务所）：中信建筑设计研究总院有限公司；肖伟、何欣然、杨青松、李伟强、李菁菁、吴航、张斯
设计时间：2015.07-2016.01
建造时间：2016.01-2016.09
用地面积：110亩
建筑面积：9400m²

Xiedian Village has been selected as one of the fourth batch of "Traditional Chinese Villages". Xiedian Village is a traditional old village, surrounded by mountain and water. It is located in Songbu Town, Macheng City, Hubei Province. Adjacent to magnificent Weidou Lake, Xiedian Village is like a wonderful lost world waiting to be discovered. Xiedian Village has a very long history. It is said that the first villagers came here were from Jiangxi Province during Ming dynasty due to one of the largest immigration movement in Chinese History. In 1950s, as Weidou Lake reservoir began to build, original village moved accordingly and formed the village we see today. Villagers keep original living and producing styles, simple but rich in cultural background.

Design Concept:
Principle of "minimal intervention": Respect the original style of the village, conform to the space texture, keep the place memory, with the principle of minimal intervention, follow the design concept of low compact, low intervention, low consumption and low maintenance. We mainly protect the houses, trees, stone arch bridges, ancient dams and city walls. and reduce the impact on the ancient village through minimal site intervention, minimal environment intervention and the most efficient utilization of environment resources to give priority to local plants and materials, play the multiple functions of materials and reduce the costs of transportation and labor, encourage villagers to participate. It is an effective way for the sustainable development of traditional villages and making profit to cultivate economic crops of ornamental value and lease the green space as a family garden and set up food and beverage facilities in the surrounding area.

Planning:
1. Feature protection: As we observed, the most important characteristic in this village is its roads and courtyards. The terrain is closely connected with mountain and water. It is conform to the terrain but not constrain to it. There are lots of traditional features we would like to keep such as stone wall base, grey wall and roof tiles. It stays in harmony with its surroundings and stands out as unique in some ways. Our design principle is to fully understand and respect the local culture and natural landscape, help the old village layout grow and stretch.
2. Planning structure: Based on the original layout, we subtly emphasize the sense of depth by embedding landscape. By connecting north and south water systems, we realize two axis of green and water. With three parts and seven districts from outside to inside, we seek to promote functional and traffic organization..
3. Environment improvement: Adhere to the principle of "minimal intervention", we hope to keep as much historic information as we can, improve the living environment of country and keep nostalgia. Its design is performed from five aspects: restoration of buildings, reorganization of roads, remediation of courtyards, enriching of vegetation and softening the revetment.

谢店村是第四批"中国传统村落"入围项目，位于湖北省东北部，地处麻城市宋埠镇，距离麻城龟峰山景区40km，毗邻山水一色的尾斗湖畔，是一处山水环抱的世外桃源。
相传，明代发生"江西填湖广，湖广填四川"的大型移民迁徙活动，几户江西群众走到此地定居，繁衍生息。20世纪50年代，随着尾斗湖水库的修建，原有老村整体搬迁上岸，形成现在的谢店古村落。村民至今保留传统的生活和生产方式，乡风淳朴，人文底蕴浓厚。

设计理念
最小干预原则：尊重村落原始风貌，顺应空间肌理，保留场所记忆，遵循低冲击、低干预、低消耗、低维护的设计理念，主要对村落古民居、古树、古石拱桥、古堤、古城墙等历史遗址进行保护，通过最少的场地介入、最低的环境干预与最高的场地环境资源利用来减少对传统村落的冲击；优先选择乡土植物以及石材等地域材料，发挥材料的多重功能与减少材料运输成本，降低人力成本。参与原则：鼓励村民、社会群体参与其中。
可持续原则：种植有观赏价值的经济类作物，对外租赁绿地作为家庭花园，在绿地周边设置餐饮休憩服务设施，有利于回笼资金，是传统村落可持续发展的有效途径。

规划设计
1. 风貌保护——阡陌纵横，院落情怀：村落接山连水，绿谷通幽；顺应地势，延绵布局；灰墙石基，瓦屋栉比；藏风聚气，浑然天成；风貌完整，特色鲜明。设计尊重地域文化及山水脉络，延续村落空间肌理。
2. 规划结构——绿水两轴，东南聚气；三带七区，环环相抱：村落东入口"长堤春柳"强化纵深感，与南北水系交汇，在后山的映衬下，形成最美绿、水两轴。三带七区由外到内，环环相扣浑然天成。
3. 环境营造——最小干预，留住乡愁：在最大限度地保护历史信息的前提下，秉持最小干预原则，改善农村人居环境，将传统村落延承性忠实地留给后人，留下乡愁。从修复建筑、梳理道路、整治院落、丰富植被、柔化驳岸五方面进行设计。

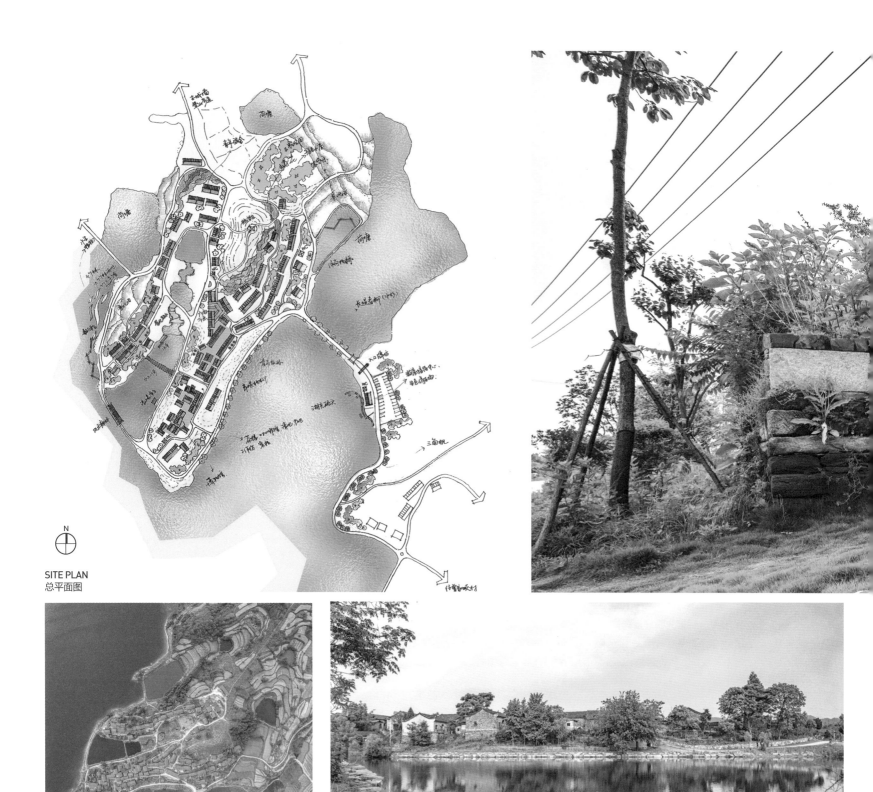

SITE PLAN
总平面图

Incremental Update Design of Gunan Street Historical Culture Districts in Yixing

宜兴市古南街历史文化街区渐进式更新设计

Location: Dingshu Town, Yixing, Jiangsu Province, China
Architect (Studio): Wang Jianguo, Bao Li, Li Haiqing, Shen Yang, Tang Peng, Yuang Chenglong, Li Jiaxiang, Wang Linyan, Liu Yiqiu, Yu Junwang, Shi Jianbo, School of architecture at Southeast University
Design: 2012-2017
Construction: 2017
Site Area: 38,423m²
Floor Area: 2,172m²

地点：中国江苏省宜兴市丁蜀镇
建筑师（事务所）：东南大学建筑学院：王建国、鲍莉、李海清、沈旸、唐芃、袁成龙、李家翔、王琳嫣、刘奕秋、余君望、施剑波
设计时间：2012-2017
建造时间：2017
用地面积：38423m²
建筑面积：2172m²

Gunan Street Historical and Cultural Districtis located in Dingshu town in Yixing City, of Li River to its west, and Mount Shu to its east. The inheritance of heritage is still alive in the district so far, maintaining a clear development vein of purple clay industry context. How to fully protect and display the cultural heritage and settlement characteristics while improving the quality of the environment and living, has become the main task of the renewal of Gunan Street.

A progressive and embedded design strategy is adopted in this demonstration project to protect the overall spatial pattern of traditional settlements while weaving the original context and texture, focusing on the renovation of critical space and environment, and promoting the function and performance of the typical traditional dwellings. There are 51 achievements developed in the program, including key technologies, patented inventions and new crafts regarding on adaptive protection and reuse of traditional settlements effectively applying in the practical renovation.

After 5 years' unremitting efforts, Gunan Street Retrofit Phase I has achieved great outcome and play an exemplary role in its neighborhood's renewal, promoting the bottom-up organic renovation in the whole district. What's more, it has attracted a lot of purple clay practitioners, reviving the traditional purple clay industry and promoting the surrounding industries. Dingshu has also been rated as one of the second characteristic town of China, attracting more and more people to experience the beautiful Chinese traditional settlement's characteristics and live transmission of purple clay culture.

古南街历史文化街区位于宜兴丁蜀镇，东依蜀山，西临蠡河。街区迄今仍有活态的紫砂工艺传承与发展脉络。如何在充分保护和展示文化底蕴和聚落特色的同时，提升环境和居住品质，是古南街更新面临的首要课题。

本工程采取的渐进式、镶嵌式更新设计策略以保护传统聚落的整体空间格局、织补原有文脉肌理为前提，对重点节点进行空间及环境整治，对传统典型民居进行功能及性能提升设计。研究开发的古建聚落适应性保护与利用成套关键技术及专利发明、新工艺等51项成果在实际更新中都得到了有效的运用。

五年的不懈努力下，古南街一期更新已初见成效，有望推动整个街区自下而上的有机更新。丁蜀也被成功授予全国第二批特色小镇，吸引越来越多的人来此感受传统聚落特色与活态传承的紫砂文化。

SITE PLAN
总平面图

Zhudian Hoffmam Kiln Culture Center 祝甸砖窑文化馆

Location: Zhudian Village, Kunshan, Jiangsu Province, China
Architect (Studio): Cui Kai, Guo Haian, China Architecture Design Group
Design: 2014-2015
Construction: 2015-2016
Site Area: 9,777m²
Floor Area: 1,650m²

地点：中国江苏省昆山市祝家甸村
建筑师（事务所）：中国建筑设计院有限公司：崔愷、郭海鞍
设计时间：2014-2015
建造时间：2015-2016
用地面积：9777m²
建筑面积：1650m²

Zhudian Village is located in Jinxi Town and known as Chen Cemetery in ancient times. This place is one of the producers of "gold bricks", which are used as the floors of the Forbidden City in Beijing. The small village is such a beautiful place whose three sides are nearby the lake and easy to reach by land and water. The villagers lived on making bricks in the old days, and they left ancient kilns which were used more than 300 years ago in the east and an old Hoffmam kiln built in 1980s in the west. However, they go out of their home town to find jobs nowadays with the demise of the ancient industry. To revive the village, we reconstruct the old Hoffmam kiln to be a kiln culture centre, which can create some routes from the entrance of the village to the ancient kilns on the other side. Therefore, the opportunities of rural development occur with the minimum fund. We just get rid of the old roofs and tiles, which are not reliable, and lie three new "safe cores" inside the building, which are steel structures. And the roofs of the "safe cores" are connected to support the new roof which are made of the old tiles and new organic plastic tiles, to simulate the inner space as it was before rebuilt. The gas with its temperature relatively constant in the kiln body is use as a heat medium to obtain good ventilation effect. All internal furniture, cabinets and floor units are modular so as to meet flexible demand.

祝家甸村位于江苏省昆山市锦溪镇，古称陈墓，是明清两代皇家宫殿所用金砖的产地。村庄三面环水，环境优美，交通便利，村子东边现存明清古砖窑20余座，村民素以烧砖为生计，村西边现存20世纪80年代大型霍夫曼砖窑一座。但是随着烧砖产业的没落，村民已经多外出打工，乡村凋敝没落。为了振兴和恢复村庄，使其继续传承原本的文化与风貌特色，故将村子口废弃的霍夫曼砖厂加以改造，利用有限的资金建设一座小的砖窑文化馆，通过村西文化馆与村东古窑之间的联络带动整个乡村的复兴。设计选取了最轻微的改造模式：将其屋顶拆除，在其内部植入新的钢结构体系，该结构体系如同一个安全核，具有安全性和稳定性。然后将三个安全核顶部相连，铺上檩条，将旧瓦和新引入的有机塑料透明瓦铺装到屋顶上，形成如同过去的斑驳效果。再利用窑体内温度相对恒定的气体作为热媒，达到冬暖夏凉的通风效果。室内全部采用组件式的家具、柜子、地板单元，以满足灵活多变的使用需求。

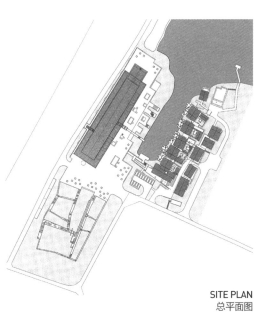

SITE PLAN
总平面图

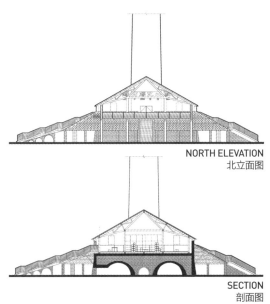

NORTH ELEVATION
北立面图

SECTION
剖面图

Macha Village Community Center

马岔村村民活动中心

Location: Macha Village, Dingjiagou Town, Huining County, Baiyin, Gansu Province, China
Architect (Studio): On-earth Architecture
Design: 2013.05-2014.04
Construction: 2014.06-2016.05
Site Area: 1,860m²
Floor Area: 648m²

地点：中国甘肃省白银市会宁县丁家沟乡马家岔村
建筑师（事务所）：土上建筑工作室
设计时间：2013.05-2014.04
建造时间：2014.06-2016.05
用地面积：1860m²
楼层面积：648m²

Coordinated by the team and the villager committee of Macha, a sub-project of village community center was launched on the site beside the crossroads to all community groups, a sloping wasteland with a good view facing the eastern valley. In response to villagers' demands, the functions of clinic of Chinese medicine, kindergarten, library, stage, shop and a multi-functional hall are combined together. It also works as a demonstration and training base of MOHURD in upgraded rammed-earth technology.

The architecture is based on team's research result on the modern rammed earth building technique and the construction experience of the previous sample house. It has efficiently used local resources and traditional construction organization method. It was the new rammed earth building technique's realization in the modern architectural design and technical demonstration and training of local craftsmen and villagers through the construction itself.

The center is located at Macha Village, Huining County, Gansu Province where is the droughty Loess Plateau's gully region with rich earth resource. The architecture has used the local traditional yard typed architecture for reference in terms of spatial arrangement. It combined the recessive foundation, setting the earth houses at different altitudes and formed a courtyard house where the opening faces a valley. We hope these earth houses can be integrated into the local landscape naturally just like earth grown from ground.

该中心为住房和城乡建设部现代夯土民居研究与示范项目中的一项重点内容，是由无止桥慈善基金会出资，团队统筹、设计并组织当地村民共同完成。其功能包括：多功能厅（含培训、展示、图书阅览）、商店、医务室和托儿所。目前，该中心已投入使用，除满足马岔村民日常公共生活服务需求外，同时也是该地区推广现代生土建造技术的培训基地。

基于团队在现代夯土建造技术方面的研究成果与前期示范民居的建设经验，项目充分利用了本地资源和当地传统的施工组织模式。这既是新型夯土建造技术在现代建筑设计中的一次实践，也是通过施工过程本身对当地工匠和村民的一次技术示范与培训。

中心所处的甘肃省会宁县马岔村为干旱的黄土高原沟壑区，土资源极其丰富。建筑在空间组合方式上借鉴了当地民居传统的合院形式，结合基地的退台现状，将若干土房子设置于不同的标高，围合出一个三合院，开口面向山谷。我们希望这几个土房子就像在地里生出的土块，可以自然地融入当地的空间景观之中。

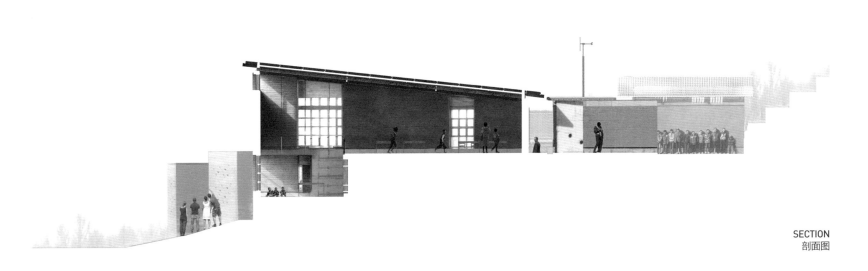

SECTION
剖面图

Contemporary Collective Living: New Forms of Affordable Housing for Relocalized Farmers in Hangzhou

乡村低收入住宅——杭州富阳东梓关回迁安置农居

Location: Dongziguan Village, Fuyang District, Hangzhou, Zhejiang Province, China
Architect (Studio): Zhejiang Greenton Architectural Design Co., Ltd.
Design: 2014
Construction: 2016
Site Area: 19277.6m²
Floor Area: 15286.98m²

地点：中国浙江省杭州市富阳区东梓关村
建筑师（事务所）：浙江绿城建筑设计有限公司
设计时间：2014
建造时间：2016
用地面积：19278m²
建筑面积：15287m²

Currently the living conditions in large part of rural China are poor. For instance in Dongziguan Village in Fuyang Hangzhou, most of farmers still live in the aged housings of disrepair. Local Government in Fuyang District of Hangzhou decided to fund an exemplary affordable housing project in Dongziguan Village aiming at improving living conditions for relocated farmers.

During the design process, architects conducted investigations and meetings to communicate with different families of the relocated farmers for first-hand information including their living habits. The project seeks to organize the buildings in the vernacular style of a courtyard typology, a traditional local morphology. The design of the courtyard makes it vary into four prototypes that are learnt from the tradition and its diversity. The prototypes could be developed into clusters, which later grow into a larger rural settlement.

The plan layout based on the common requirements from the relocated farmers tries to balance the traditional rural life-style and high-quality modern living condition. The design of the housings is not a carbon-copy of the local historic buildings, but abstracts and refines the features of traditional local architecture with contemporary understandings, and then incorporates them into the design of new housings.

东梓关村中部分原住民仍然居住在年久失修的历史建筑中，为了改善居住与生活条件，当地政府决定采用政府代建的模式进行回迁安置，打造具有一定推广性的新农居示范区。

建筑师采用实地调研、座谈测绘的工作方式切入设计，寻求真实的第一手资料。本项目在规划上从传统肌理的院落空间基本单元出发，遵循从单元生成组团，再由组团演变成村落的生长逻辑，通过四种基本单元的组合再现传统聚落的多样性。

平面功能空间从农民真实需求出发，回归生活本源，寻找一种介于传统民居和城市化居住模式之间的状态。立面上没有拘泥于传统地域民居的造型符号，而是对其提取解析并加以抽象，外实内虚的界面处理塑造传统江南民居的神韵和意境。

3RD FLOOR PLAN
三层平面图

2ND FLOOR PLAN
二层平面图

REGIONAL MASTER PLAN
区域总平面图

1ST FLOOR PLAN
一层平面图

Renovation of Countryside Buildings of Xibang Village

西浜村农房改造工程

Location: Xibang Village, Kunshan, Jiangsu Province, China
Architect (Studio): Cui Kai, Guo Haian, China Architecture Design Group
Design: 2014-2015
Construction: 2015-2016
Site Area: 2,775m²
Floor Area: 1,643m²

地点：中国江苏省昆山市西浜村
建筑师（事务所）：中国建筑设计院有限公司：崔愷、郭海鞍
设计时间：2014-2015
建造时间：2015-2016
用地面积：2775m²
建筑面积：1643m²

Nestled against the Chuodun Hill and facing the Yangcheng Lake, Xibang Village is the birthplace of Kun Opera, one of the oldest opera type in China. To restore energy of the village by reviving its cultural atmosphere, four courtyards enclosed by white walls and bamboo screens are renovated as a school of Kun Opera. Stage is set adjacent to the river course and two layers of corridors connect the spaces. It becomes a comfort space for opera studying.

西浜村位于阳澄湖畔、绰墩山北，是昆曲文化的发祥地。如今村庄已经破落，为了恢复乡村中的昆曲文化，带动乡村复兴，将村中的四座老宅院修复改造，设计成昆曲学社。通过粉墙和竹墙形成梅、兰、竹、菊四院，结合水系设计了戏台。通过两层游廊的穿插，形成一个空间丰富、光影交错的昆曲研习场所。

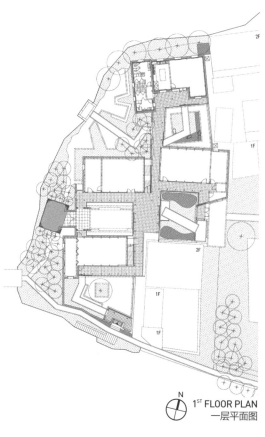

1ST FLOOR PLAN
一层平面图

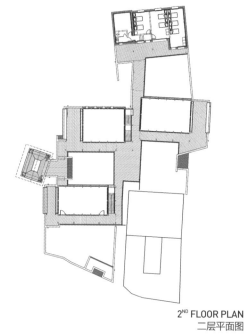

2ND FLOOR PLAN
二层平面图

ELEVATION
立面图

THE-Studio
(Tsinghua Eco Studio)

清控人居科技示范楼

Location: Gui'an District, Guizhou Province, China
Architect (Studio): SUP Atelier, Beijing Tsinghua Urban Planning & Design Institute
Design: 2015.02-2015.04
Construction: 2015.06
Site Area: 1,826m²
Floor Area: 701m²

地点：中国贵州省贵安新区
建筑师（事务所）：北京清华同衡规划设计研究院有限公司 SUP 素朴建筑工作室
设计时间：2015.02-2015.04
建造时间：2015.06
用地面积：1826m²
建筑面积：701m²

THE-Studio (Tsinghua Eco Studio) is the first demonstrating and experimental Nearly Zero Energy Building of the Gui'an Innovation Park in southwest China. Located at the park's entrance, it connects the constant ecological sponge city landscape, providing an extremely fascinating scenery for both users and visitors. Moreover, its two floors above ground can serve as large-space exhibition, VIP reception & conference and daily office, while one single underground floor can act as equipment and storage area.

THE-Studio is both a practical demonstration case which deeply integrates sustainable design strategies with ecological technologies, and an experimental platform for sustainable architecture based on the Moderate Zone in southwest China, aiming to examine whether the detailed design methods and building technologies can truly suit to the local conditions of climate, culture and even economy.

Oriented to the specific project positioning, the design team has developed a multi-system integrated design strategy from the very early design phase to minimize adverse impact upon local ecosystem and maximize indoor comfort and energy efficiency.

After the building was completed, a series of field measuring and monitoring for thermal, humid, ventilation and luminous environments are carried out and further analyzed. The outcomes verify that detailed design methods and building technologies of THE-Studio at the early design phases were both effective and appropriate during the construction and operation processes, which can be the references to similar sustainable buildings in the Moderate Zone of southwest China.

清控人居科技示范楼是在贵安新区政府支持下，清控人居建设集团与英国 BRE 机构合作的示范项目，旨在建成符合 BREEAM 标准的近零能耗示范实验建筑。建筑位于贵安新区生态文明创新园内，是立足于中国西南温和地区的可持续策略实验平台，以检验各项设计方法与技术措施在当地气候、文化甚至经济条件下的可行性。本项目采用多系统整合的建构方式，主要体现于三个层面：多系统并行建造、乡土文化与可持续技术的整合。

整体建筑由木结构系统、轻钢箱体系统、设备系统、外表皮系统四部分并行建造而成，均在工厂预制并现场吊装，不仅有效节省时间与资源成本，更为后续实验平台中设备系统的增补、操作与检修提供便利。

本项目将主动式节能技术与建筑布局、空间形态、自然通风采光等内容有机结合，形成了独具特色的室内室外空间。同时大量应用可再生材料，并鼓励采用当地特有的乡土材料与工艺，在建筑全生命周期中有效降低建筑碳足迹，并创造出独特的建筑本土表现力。

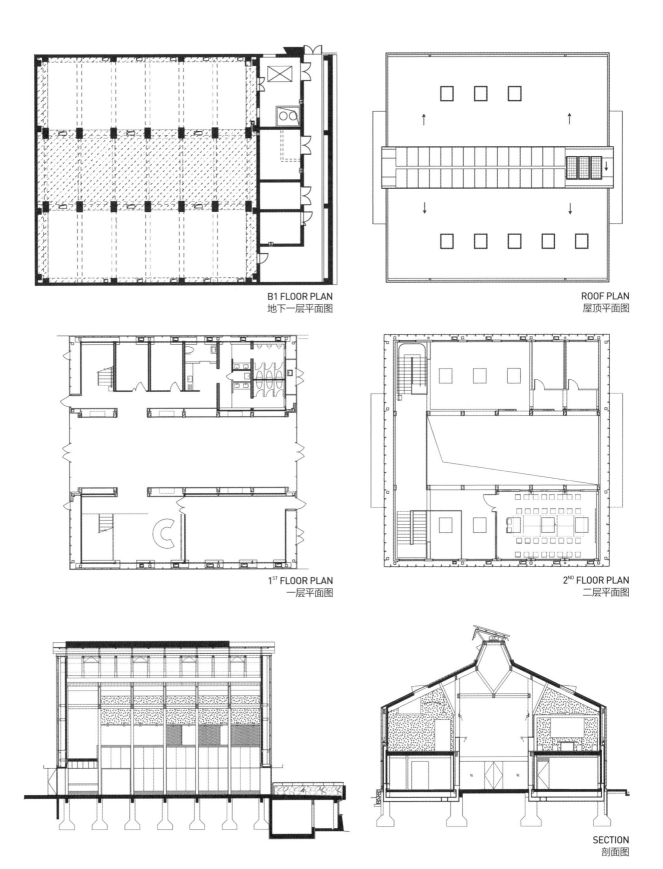

Renovation of Wencun Village 文村村新建民居

Location: FuYang, Hangzhou, Zhejiang Province, China
Architect (Studio): Amateur Architecture Studio
Design: 2013.01-2015.06
Construction: 2014.07-2015.10
Site Area: 48,905m²
Floor Area: 37,680m²

地点：中国浙江省杭州市富阳
建筑师（事务所）：业余建筑工作室
设计时间：2013.01-2015.06
建造时间：2014.07-2015.10
用地面积：48905m²
建筑面积：7590m²

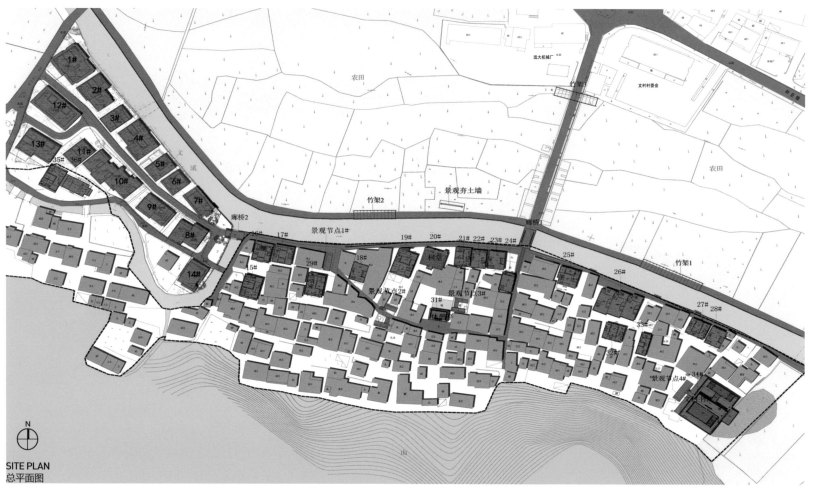

SITE PLAN
总平面图

- 拆除建筑
- 改造建筑
- 保护及修缮建筑
- 新建及原址替换建筑
- 新建景观节点

ANALYSIS DIAGRAM
分析图

Hani Nationality Mushroom House Retrofit Experiment in Azheke Village, Yuanyang County

元阳县阿者科村哈尼族蘑菇房改造实验

Location: Honghe Hani and Yi Autonomous Prefecture, Yunnan Province, China
Architect (Studio): Zhu Liangwen, Chen Xiaoli, Cheng Haifan, Faculty of Architecture and City Planing, KUST; Architecture Institute of Vulgar Autochthonous, KUST
Design: 2015.05
Construction: 2015.07-2015.11
Site Area: 120m²
Floor Area: 245m²

地点：中国云南省红河哈尼族彝族自治州
建筑师（事务所）：昆明理工大学建筑与城市规划学院、昆明本土建筑设计研究所有限公司：朱良文、陈晓丽、程海帆
设计时间：2015.05
建造时间：2015.07-2015.11
用地面积：120m²
建筑面积：245m²

Azheke has been included in the third batch of the Traditional Chinese Villages Catalog, and it is also the traditional village of the Hani nationality that preserves the most intact style and feature with rather complete inheritance of intangible culture within the world landscape cultural heritage area of Honghe Hani Rice Terraces.

The folk house renovation project is located in the village of Azheke, and it chooses the comparatively typical Hani traditional folk houses. Based on the reservation of the traditional spatial textures and architectural styles of the village, the functions and the internal environment of the traditional dwellings have been improved and upgraded: the 1st floor has adopted the method of locally down digging to utilize the bottom space rationally and transform it into a bar and a space with auxiliary functions; meanwhile, kitchen and bathroom have been added to enhance its functions. The 2nd floor has been divided into the spaces for exhibition and living to improve the previous situation of unclear space division. The 3rd floor has been adapted into a simple living space to ease the shortage of living space. Throughout the renovation process, local construction materials and building technology have been selected, and the renovation style kept as simple as possible so as to play an universal demonstration role in the renovation of traditional residences, change the thoughts of local villagers demolition and reconstruction, explore new methods for the traditional residential renovation of the Hani nationality and the new ideas on protection and development of the traditional villages, and further inherit and develop the traditional residential cultural heritage of the Hani nationality.

阿者科村被列入第三批中国传统村落名录，是红河哈尼梯田世界景观文化遗产区内风貌保存最为完好、非物质文化传承较为完整的传统哈尼族村落。

民居改造项目位于阿者科村内，选取较为典型的哈尼族传统民居。在保留村落传统空间肌理和建筑风貌的基础之上，对传统民居使用功能及内部环境进行改造提升：一层采用局部下挖的方式，合理利用底层空间，改造成酒吧及附属功能空间，同时增设厨房及卫生间，使用功能更加完善；二层分割为展览及居住空间，改善以往空间分区不明的现象；三层改造成为简易的居住空间，缓解居住空间紧缺的问题。在整个改造过程中均选用本土建筑材料及建造工艺，改造风格尽量朴实，以期对当地民居起到普适性的改造示范作用，改变村民拆旧建新的想法，探索哈尼族传统民居改造以及传统村落保护与发展的新思路，进而使得哈尼族传统民居文化遗产得以延续发展。

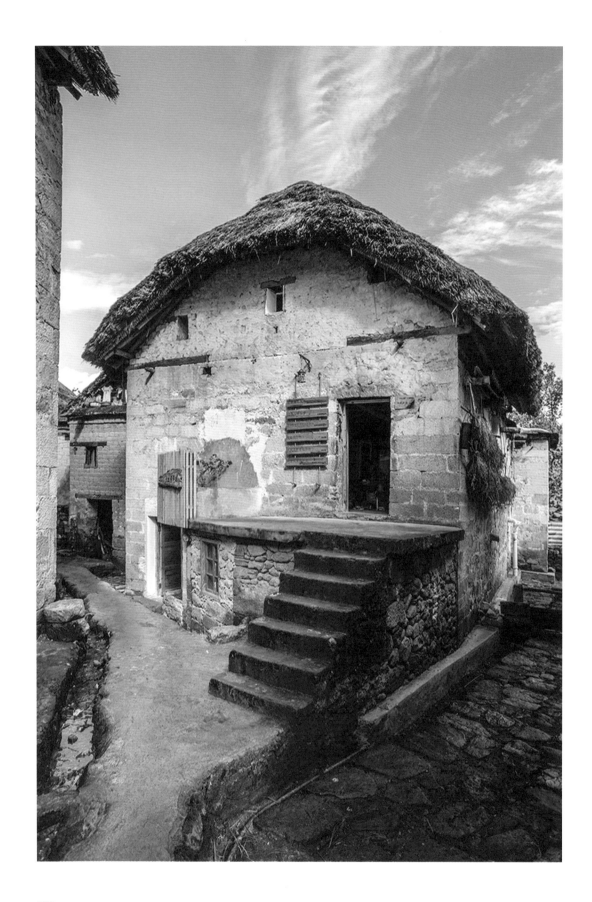
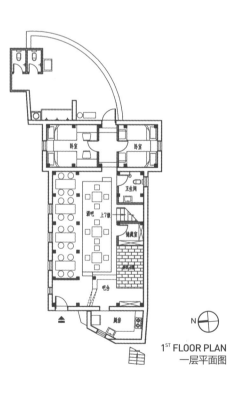

1ST FLOOR PLAN
一层平面图

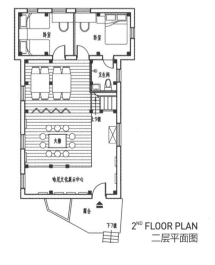
2ND FLOOR PLAN
二层平面图

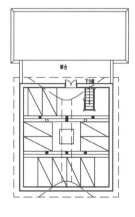
3RD FLOOR PLAN
三层平面图

SECTION
剖面图

A Rural Shop at Huashu Village, Nanjing

南京桦墅村口
乡村铺子

Location: Zhouchong, Huashu Village, Qixia District, Nanjing, Jiangsu Province, China
Architect (Studio): Atelier Zhou Ling, Institute of Architecture Design & Planning Co., Ltd, Nanjing University
Design: 2014.09
Construction: 2015.06
Site Area: 668m²
Floor Area: 150m²

地点：中国江苏省南京市栖霞区桦墅村周冲
建筑师（事务所）：南京大学建筑规划设计研究院有限公司周凌工作室
设计时间：2014.09
建造时间：2015.06
用地面积：668m²
建筑面积：150m²

The Rural Shop was originally an old house located at entrance of the village. The design attempted to modify it into a public activity space so as to restore traditional Chinese "Village Entrance" space. Traditional layout and construction methodology have been used to complete the rebuilding of village entrance tradition and reshape the publicity of the village entrance.

The original village entrance building was an ordinary three-bay house, 11 meters wide, 7 meters deep. According to the original house site, we built a 6 x 4.8 meters public gallery for villagers and passing visitors to rest and communicate, a village entrance landscape square and trees are provided outside.

We chose the traditional Chinese wood structure as roof frame, hoping to make the form back to the traditional local character, reflecting the construction of the local system. The design also referred to public buildings in the South Anhui and Wuyuan villages, such as pavilions, covered bridges, etc. These rural construction types had been widespread in the past Jiangnan villages. Thanks to traffic inconvenience and other reasons, these rural buildings were retained in southern Anhui and Wuyuan. Outdoor corridors extended into the interior, and indoor and outdoor structures were consistent. Today local managers in Nanjing have participated in the rural shop. In the weekdays traditional agricultural products will be sold and in display and in the weekends it will be a rural market in combination with the village's entrance square for villagers and tourists buying and selling goods.

乡村铺子位于村口，试图恢复中国传统乡村的"村口"空间，把村口一栋老房子改造成公共活动的场所，希望用传统的布局和工法完成对村口传统的重建，重新塑造村口的公共性。

原村口建筑是普通的一层三开间房屋，面宽11m，进深7m。设计按照宅基地原址重建，并在外侧结合村口景观广场和大树，加建一个6m x 4.8m的公共性廊子供村民和过往游人休息交流。

屋架采取传统中国的木构做法，希望尽量使它的形式能回溯传统地方性格，体现江南本土的建造体系。设计参考了皖南和婺源乡村的一些公共建筑形式，如路亭、廊桥等，这些乡村建筑类型是过去江南乡村曾经普遍存在的，只不过在皖南和婺源这些地方，因交通不便等原因这些乡村建筑被保留了下来，基本可以把它们理解成江南乡村过去普遍的样子。室外的廊子延伸到室内，室内外部分结构是一致的。如今乡村铺子已落实南京当地的经营者进驻，平日里进行传统农产品的展示与售卖，周末时结合村口的景观广场会成为买卖物品的乡村集市。

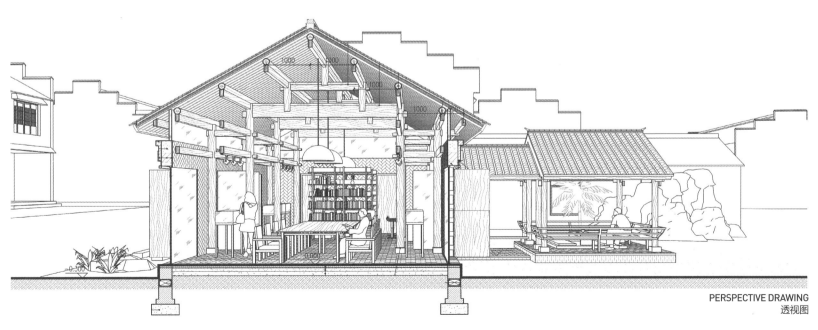

PERSPECTIVE DRAWING
透视图

277

Oriental Rural Life Pavilion, Huishan, Wuxi

无锡惠山东方田园生活馆

Location: Yangshan, Huishan District, Wuxi, Jiangsu Province, China
Architect (Studio): Qian Qiang, United Design Group Co. Ltd., Shanghai
Design: 2013
Construction: 2013
Site Area: 1,500m²
Floor Area: 1,337m²

地点：中国江苏省无锡市惠山区阳山镇
建筑师（事务所）：上海联创建筑设计有限公司；钱强
设计时间：2013
建造时间：2013
用地面积：1500m²
建筑面积：1337m²

The project is located in the "peach town of China"—Yangshan, Wuxi City. It is the first countryside complex project in China. Based on the background of creating the overall environment of "beautiful countryside" and the core target of "rural lifestyles", Oriental Garden tries to execute the concept of protecting ecology and environment.

The planning design principles are to respect natural ecology and the historic memories of Shifang Village, keep the original spatial shape and scale of existing villages to the full extent, and maximize the use of the existing road and site system.

The project design shall be rooted in rich natural and humanistic resources, seeking the harmonic coexistence between the new buildings and existing environment, and creating buildings that belong to the specific site and time, so as to provide proper spatial shape for the new lifestyle of "modern countryside".

Space constitution comes from the traditional architectural space prototype — garden space, which is enclosed by solid wall on east and west with open space formed north to south. The façade design of the east and west wall is featured with a local symbolic element—peach blossom, setting off by contrast the theme of "peach culture", and bringing along strong regional culture ambience. With the cladding of the hollow aluminium board with peach blossom pattern applied, the building appearance looks more variable. Simultaneously, the building is given with dual nature: confident yet delicate at daytime, romantic and fantastic at night. The building responses to the tradition and local culture while at the same time reveals characteristics and elements of times.

该项目位于"中国水蜜桃之乡"无锡市惠山区阳山镇，是国内首个田园综合体项目，以"美丽乡村"的大环境营造为背景，以"田园生活"为目标核心，贯穿生态与环保的理念。

规划设计原则是尊重自然生态和"拾房村"的历史记忆，最大程度保持原有村落尺度及空间形态，最大限度利用原有道路及场地系统。保护更新原始村落，延续田园生活场景。

项目设计根植于丰富的自然及人文资源，寻求新建筑与原有环境的协调共生，塑造根植于此时此地的建筑，为"当代田园"这一新型生活方式提供与之相适应的空间形态。

空间构成源于传统建筑的空间原型——庭园空间，围合庭园的建筑南北通透、东西实墙建筑的东西侧外墙设计提取了地域的象征元素——桃花，烘托"桃文化"主题，带来浓郁的地域文化气息。通过桃花图案的镂空铝板的运用，使建筑表情丰富且具有双重性：白天稳重、细腻，夜晚浪漫、梦幻。建筑在对话传统和地域文化的同时，又体现了时代特征和元素。

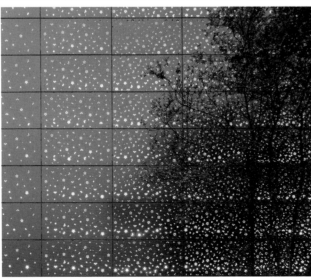

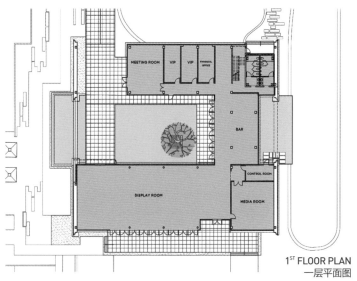

1ST FLOOR PLAN
一层平面图

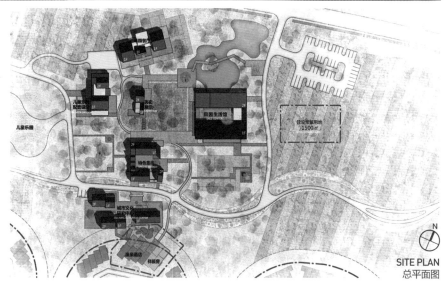

SITE PLAN
总平面图

Comprehensive Building and Academic Center of FAST Engineering Observation Base

FAST 工程观测基地综合楼和学术交流支撑平台及附属设施工程

Location: Guizhou Province, China
Architect (Studio): China IPPR International Engineering Co., Ltd.
Design: 2013.06-2014.05
Construction: 2014.02-2016.10
Site Area: 5,860m²
Floor Area: 2,680m²

地点：中国贵州省
建筑师（事务所）：中国中元国际工程有限公司
设计时间：2013.06-2014.05
建造时间：2014.02-2016.10
用地面积：5860m²
建筑面积：2680m²

The site is located at Buyi and Miao autonomous prefecture of Guizhou Province. This project is an observation base serving for FAST (Five Hundred Meters Aperture Spherical Radio Telescope), and it integrates academic exchange, office, scientific experiments and accommodation.

Since ancient times, the unity of heaven and human has been always the theme of Chinese thinkers, the harmonious coexistence of human and nature is the perfect pursuit of architectural design. The site is located at a valley with marvellous views, there is pleasant landscape surrounding the site, with mountains and amounts of pines. The surrounding geological conditions are naked Karst and dense ground-vegetation.

The layout takes advantage of the topography, and physical relief has been fully utilized to reduce earthwork volume and save investment. Also, using of environmental friendly and energy saving technologies and easy-construction materials highlights the respects to Mother Nature. At the same time, nature becomes the main architectural experience thing. Local materials create the continuous atmosphere, however the modern design methods show the fresh identity of the building by studying and referencing spatial pattern of Guizhou folk houses and villages layout.

项目位于中国贵州省黔南布依族苗族自治州，是为 FAST（500m 口径球面射电望远镜）服务的集学术交流、办公、科学实验、住宿于一体的观测基地。

自古以来，中国人就尊崇"天人合一"的至高境界，人与自然的和谐相处是建筑设计的完美追求。基地选址于一处风景优美的山谷，处于群山密林的环抱之中。用地四周覆盖为大片浓密的松林，地面由裸露的喀斯特地貌岩石和丰富的绿草植被组成。

建筑布局依山就势，充分利用自然地形，减少土方量，节约投资。并采用环保、节能、便于施工的材料，对环境给予最大限度的保护。同时追求建筑融入自然，以现代的设计手法、绿色节能的材料去诠释地域传统建筑的精髓，对于贵州民居的空间形式以及村落的布局结构研究借鉴，力求塑造出自然清新的建筑精品。

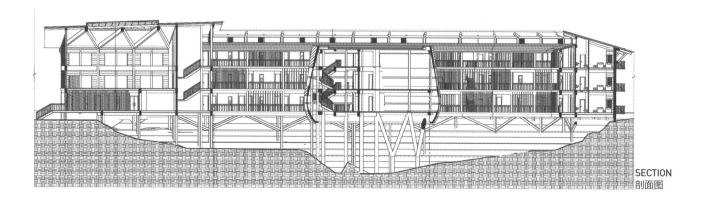

SECTION
剖面图

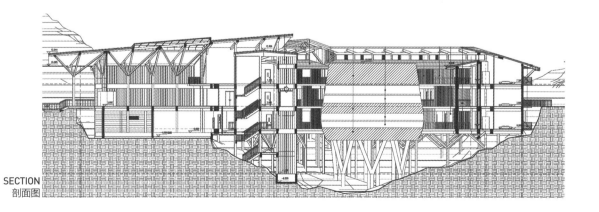

SECTION
剖面图

The Yinlu Tea House of Lianhuadang, Yixing

隐庐莲舍：宜兴莲花荡农场茶室

Location: Dingshu, Yixing, Jiangsu Province, China
Architect (Studio): Tang Peng, School of Architecture at Southeast University
Design: 2015.05-2015.11
Construction: 2015.11-2016.05
Site Area: 392m²
Floor Area: 392m²

地点：中国江苏省宜兴市丁蜀
建筑师（事务所）：东南大学建筑学院：唐芃
设计时间：2015.05-2015.11
建造时间：2015.11-2016.05
用地面积：392m²
建筑面积：392m²

The project is located in Lianhuadang Farm, Yixing. The first issue is how to build a teahouse to meet the local characteristics of the outskirts of the city. Though there is endless rice fields, reed marshes and mountains, it's not always quiet in site – because there are high-speed trains roaring past the rail line nearby regularly. The water is a white stretch, joining the sky at the horizon. The designers have proposed to arrange three tea rooms in a line around the joint of water and fields, but not to interrupt the natural scenery. Each of three tea rooms represents a distinct theme. People could meditate in the courtyards and free themselves.

The second issue is how to build a series of view-observing facilities to fit the landscape and the teahouse. The construction of a series of bamboo-made facilities gives the farm a new look. It is an ideal place for local residents to enjoy themselves.

The designers studied the local building materials and construction methods in Yixing, and used materials of bamboos, woods, potteries, soil, and so on, in design. They participated in the whole construction, including arranging landscapes, selecting furniture, and making decorations.

如何在城市近郊的乡村风景中建设一座符合当地特色的茶室，是莲花荡农场给我们出的第一个题目。一望无际的稻田，芦苇荡和远山是一贯的农场风景，然而基地给予我们的是并不持久的安静，与地平线平行的高架线路上定时有高铁呼啸而过。水天一色，古往今来。在这样一片现代的乡村风景中，我们选择在田地与水体交界处，设计一座横向展开的一字形建筑，三间茶室，三种意象，一处内院，与山水同坐，观天地人心。

在广阔的田野上，如何建造起一系列观景设施来配合农场与茶室的四季景观，是莲花荡农场给我们出的第二个题目。一系列竹构设施的建造，赋予农场新的表情。为此莲花荡农场也成为一个周边居民能够自由赏游的场所。

建造上，我们尝试研究宜兴当地特有的建筑材料和建造方式，将竹、木、陶、土等材料悉数表达。并在实际建造过程中，由建筑师直接参与从局部施工、景观布置，到室内装修，以及家具选择、器物摆设的全部过程。与世界的温柔相望，是我们对莲花荡农场的回答。

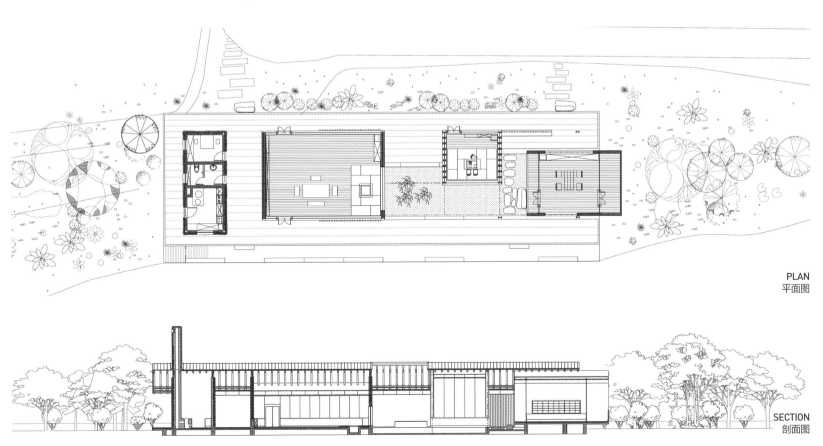

PLAN
平面图

SECTION
剖面图

Lounge Bridge Renovation

驿道廊桥改造

Location: Bali Village, Fuping County, Hebei Province, China
Architect (Studio): Atelier Heimat
Design: 2016.05-2016.07
Construction: 2016.09-2017.01
Site Area: 320m²
Floor Area: 193m²

地点：中国河北省阜平县八里庄村
建筑师（事务所）：合木建筑
设计时间：2016.05-2016.07
建造时间：2016.09-2017.01
用地面积：320m²
建筑面积：193m²

The village is divided by two rivers into three blocks. The east and west sides are connected by a bridge built in 2013. The bridge was constructed with concrete and had metal tube handrail on the two sides, 50 meters long and 4 meters wide, but because of the high density of housing and narrow roads, the cars currently cannot pass through the bridge.

The goal is to convert the bridge into a shared space for public outdoor activities. It highlights visual experience when passing through. The finished product of timber is used. All the timbers have the same sections. They are interwoven with each other through connections and inserts. Both bridge floor and handrails are using this kind of timber material.

The uniform timber can be joined to form various sections. The designer make the scale of bridge suitable for pedestrians. This approach provides increased opportunities for people to stop and sit down for sightseeing. Playful structures and canopies at the center of the bridge enable various kinds of activities for villagers, like gatherings for village affairs, playing games for kids, or simply enjoying the sunshine for seniors.

The construction work is much easier by using uniformed timber components with no cast-in-situ material processing and saving manpower. The construction was successfully built by only three skilled local woodworkers.

两条内穿河流将村庄划分为三个居住片区。一座建于2013年的桥连接东西两岸，总长度约50m，宽度为4m；现状桥面为混凝土铺设，金属圆管栏杆，但由于两端民房密集，道路狭窄，因此目前无法通车。

对此桥的改造的主要意图是将其转化为村民的公共活动空间，同时丰富通行的观景感受。设计上只通过一种断面的成品木料进行穿插、组合、连接，形成独立的结构体和空间围合，将桥面和栏杆的改造一朝搞定。

同一种木构件组合成不同的剖面空间，不同的剖面空间再进行组合，将桥划分为段落。把桥从车辆的通行尺度转变为人的行为尺度：可以凭栏眺望，也可依靠而坐，桥中央有可绕玩的构件，也有以免日晒雨淋的遮蔽处。可供老人休息晒太阳、小孩玩耍、村民相聚商讨公共事务。统一构件尺度，简化构造方式，使得施工操作简单易行，减少材料的现场加工，减少人力。

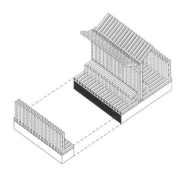
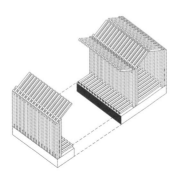
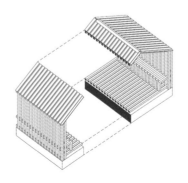
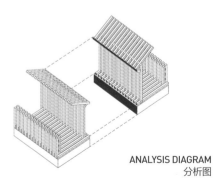

ANALYSIS DIAGRAM
分析图

SITE PLAN
总平面图

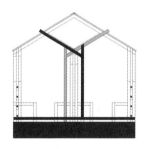
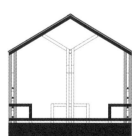
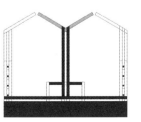

SECTION
剖面图

Wooden Bridge 木桥听雨

Location: Chujia village, Laizhou, Shangdong province, China
Architect (Studio): WangLei Architects
Design: 2015.04
Construction: 2015.09
Site Area: 200m²
Floor Area: 150m²

地点：中国山东省莱州市驿道镇初家村
建筑师（事务所）：王磊联合建筑师事务所
设计时间：2015.04
建造时间：2015.09
用地面积：200m²
建筑面积：150m²

The main body of the building sits across the river, smartly using height difference to create rich spatial variations. The north and south sides of the building are flat open wooden windows. The ground floor, transparent and full of sense of spacing, is mainly used for leisure tea and beverage. Taking advantage of the high difference between the attic and the main body, the designers create a 260 degree sightseeing platform for more open views.

All the building is in wooden structure. The bottom part of the bridge is supported by over 120 logs and 10 lumps of woods. They are connected by tenon-and-mortise work, and create a sense of order and show the beauty of the structure. The upper part is wooden frame, being applied with anti-corrosive processing wooden boards outside. The wood texture is harmoniously mixed with the surrounding natural environment. The traditional local grass roof has good thermal insulation effect and coordinates with architecture and landscape of the village.

建筑主体横跨河道，巧妙地运用高差创造出丰富的空间变化。南北两侧均为平开木窗，首层的主要功能为休闲茶饮，通透又充满空间感；阁楼部分与主体产生高低错落的关系，用作260度观景平台，使观景的视野更加开阔。建筑全部为木结构，底部的支撑采用120余根圆木以及10余根方木通过榫卯方式连接，形成秩序感，表现出建筑的结构之美。上部采用木屋架外包木板，木板做防腐处理，木头的质感与周围的自然环境融合。屋顶采用当地传统山草屋面做法，起到良好的保温隔热作用，又和村落建筑风貌协调。

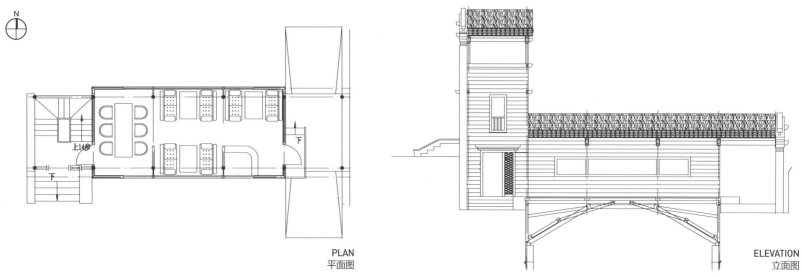

PLAN
平面图

ELEVATION
立面图

Renovation Project of Grandma's Yard in Huangshandian Village

黄山店姥姥家民宿改造项目

Location: Huanshandian Village, Fangshan District, Beijing, China
Architect (Studio): Evolution Design
Design: 2016.04
Construction: 2016.09
Site Area: 720m²
Floor Area: 305m²

地点：中国北京市房山区黄山店村
建筑师（事务所）：空间进化（北京）建筑设计研究院有限公司
设计时间：2016.04
建造时间：2016.09
用地面积：720m²
建筑面积：305m²

In Huangshandian Village Grandma's Yard renovation project, with little budget and under a very short time frame, Evolution Design turned obsolete village in the countryside into a perfect make-over by saving the texture and special relationship between the original farm house building and the surrounding courtyards, merging with the arteries and veins of the village. With the pursuit of preserving the countryside memory and making adaptations as needed to the surrounding environment, they persists in the principle of originating "outside" of tradition, yet coexists with tradition, allowing it to extend and cherish in a more respectful way.

Different from many individual design projects seen in the past, Evolution Design attempts to explore a brand-new business model to resolve the social issues commonly found in China's rural area today. Traditional villages left a large number of vacant farm houses in unlivable conditions and infrastructure fell way behind. Most young adults move to the city for work, leaving behind large number of seniors and unattended young children. As most of these villages tend to be far from cities, local farm products are difficult to be transported out, exacerbating the poverty issue. On the other hand, city residents increasingly wish to spend the weekend in the countryside to relax, getting rid of air pollution and work pressure, but with difficulties to find good quality footholds to stay. Our business model is to look for vacated houses in the countryside villages, convert these places into design-quality living space under a short time span and with good cost management. While training local farmers to provide warm room services, our food specialists would develop dining menus based on local agricultural products available, and finally, internet platform is used for marketing and sales. 75% sales revenue belongs to local farmer, and 25% covers our platform's operating cost. We have completed renovation for almost 40 courtyards in the past year, controlling cost under 60 USD per square feet (all expenses inclusive); our renovation job complete in less than 2 months on average. Using local materials and labours greatly lowered operating cost in the later phase, and sharing most of the income to the local farmers makes the model stable in nature, while attracting more farmers willing to provide vacant houses and to work together with us. The sales of courtyards having been subject to open-opration are booming, and the income of farmer housekeepers has exceeded their income from working in cities. This has driven quite some number of young people returning to the village to join us, indirectly contributing to resolv a long-standing social issue.

在隐居乡里黄山店村姥姥家民宿改造的项目中，空间进化在时间和造价都十分有限的条件下让破旧的农村老宅完成了一次华丽的变身。对于传统的"打破"并没有将传统消灭，而是让传统以一种更有尊严的方式得以延续。我们的设计原则是尽量就地取材，节约成本的同时也自然地体现出当地文化。

与以往传统的单一建筑项目不同，我们试图探寻一种崭新的商业模型来解决当前中国农村普遍存在的社会问题。传统的村落年久失修，大量空置农舍破败不堪，基础设施落后。村中的青壮年大多进入城市打工谋生，留下村中大量老年及儿童无人照料。由于这些村落大多地域偏远，农产品不便运出，进一步加剧了当地的贫困。而另一方面，城市人又受生活压力和空气污染等因素所困，希望周末能到郊区放松身心，但苦于找不到有品质保障的落脚点。我们的模型是在村子里面找到被空置的房屋，通过设计，用最快的时间和最低的成本达到一定的居住品质，之后培训当地的农民做客房服务，并由专门的团队根据当地的农作物开发出餐饮菜单，最后利用互联网平台进行销售。销售的75%归当地农民，25%覆盖服务平台的成本。我们在一年的时间里已经完成了将近四十个院落的改造，成本基本控制在了60美元/平方英尺（涵盖了场地清理、建筑改造、室内硬装、家具陈设、灯具、机电、上下水等所有费用），改造时间大概不到两个月。利用当地材料和劳动力极大地降低了后期的运营成本，而将收入的大部分分给当地农民又使得商业模型变得稳定，同时吸引了越来越多的农民愿意将手中不用的房屋提供给我们进行合作。已经开放经营的院落销售情况火爆，农民管家的收入远远超过了他们进入城市打工的收入。这使得很多本村的青壮年劳动力开始返回村子加入我们，从而间接地解决了很多社会问题。

PLAN
平面图

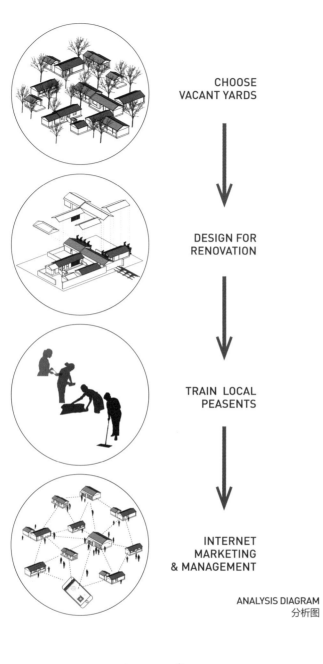

CHOOSE VACANT YARDS

DESIGN FOR RENOVATION

TRAIN LOCAL PEASENTS

INTERNET MARKETING & MANAGEMENT

ANALYSIS DIAGRAM
分析图

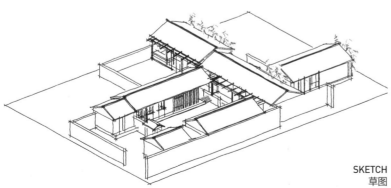

SKETCH
草图

301

Mendeng Village Community Center

门等村村民活动中心

Location: Mendeng Village, Lianhua Town, Guilin, Guangxi, China
Architect (Studio): On-earth Architecture
Design: 2015.06-2015.10
Construction: 2015.08-2015.10
Site Area: 257m²
Floor Area: 213m²

地点：中国广西壮族自治区桂林市莲花镇门等村
建筑师（事务所）：土上建筑工作室
设计时间：2015.06-2015.10
建造时间：2015.08-2015.10
用地面积：257m²
建筑面积：213m²

Aiming at the "abandoned dwellings" phenomena frequently happening in traditional villages of China, the designer teams launched a demonstration project based on the Mendeng village. According to villagers' demands for public services and activities, an abandoned courtyard with a two-story house and three-story house is selected and transformed to the village center.

Prior to the renovation, the site is formed with two traditional houses which are abandoned by the residents. Through conversation with the villagers, all agreed to transform the houses into Mendeng Village Public Activity Center which also plays the role of training and demonstration avenue for the local artefacts and culture. Base on the evaluation of the structural safety and heritage values at early-stage, our renovating design strategy has focused on five perspectives: 1. Preservation of the exterior historical characteristics, promotion of the interior space and the environmental quality; 2. Reinforcement and restoration of structural components, promoting the sustainability of the house; 3. Improvement in spatial arrangement, satisfying the multi-functional needs; 4. Through delicate detailed design, realizing the organic combination of tradition and modernity; 5. Improvement of natural ventilation and natural daylighting, upgrading interior comfort. Renovated space accommodates villagers for public events such as exhibition, assembly, training, reading, chess and cards, tea tasting, etc. This project has been undertaken for three months, the team was on site for designing, launching and guiding local villagers to succeed the construction together.

设计团队针对传统村落中较为普遍的民居空废化现象，以门等村一座废弃的民居院落为对象，通过与村民互动将其改造设计成为村民活动中心。同时也作为该地区传统营建工艺与文化的展示和培训基地。改造前的院落，由两栋传统民居建筑组成，已被原住户废弃。基于前期的结构安全和传统价值评估，设计团队采用的改造设计策略可概括为五个方面：1.保护建筑外部传统风貌，提升室内空间及环境品质；2.加固或修缮结构构件，提升房屋耐久性能；3.优化空间布局，满足多种功能需要；4.通过精细化设计，实现传统与现代的有机结合；5.改善室内自然通风和自然采光，提升室内舒适度。改造后的空间可满足展示、集会、培训、图书阅览、棋牌、品茶休闲等村民的公共活动。该项目历时三个月，由团队全程驻守现场进行现场设计，并发动和指导当地村民共同实施完成。

EAST ELEVATION
东立面图

SECTION
剖面图

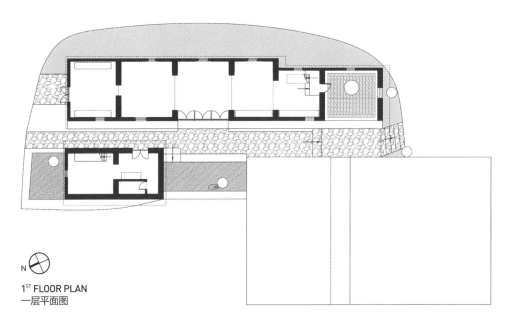

1ST FLOOR PLAN
一层平面图

Chinese Architects

中国建筑师

Chinese architecture is undergoing great changes. It is an inevitable product of social transformation and development, but most of all it is Chinese architects' efforts that count. The country's economic growth brings unprecedented amount of construction, and provides platforms for them to fulfill themselves. They follow their callings, dedicated to promoting the country's rural and urban construction and living environment. They are willing to hold on to their dreams and not daunted by high intensity of work. In face of challenges, they are making unremitting efforts to find out solutions. They traceback and rethink about the history of Chinese architecture and are always pursuing progress and make breakthroughs.

Today Chinese architects have outstretched to many fields that they have not tried before. They are making renovations in their profession practiceand seek to discover the very roots of architecture; they believe in giving back and contribute to society; they get back to the world of life to find inspirations and apply them to their designs.

Discover the Very Root of Architecture

High speed development has brought about problems of homogeneous cities and buildings. Chinese architects are rethinking if they have overemphasized speed and if there's something important that they might have missed. They pursue enduring truth in architecture and are wondering what is next for Chinese cities and buildings. In their diversified experimental practices, they compromise the merits of Chinese and western styles by constant learning and innovating, and finally find their own way. They believe it is important to have their own voice instead of following others' steps, and to discover the very roots of architecture: nature and region, history and culture, art and technique.

Contribute to Society

Architect shoulders social responsibilities. They improve living environment not only for ordinary people but also for neglected groups, as society is made up of different individuals. In metropolitan cities, they renew urban marginal areas and upgrade survival surroundings for the disadvantaged groups. In rural area, they retrieve the declining civilization and revitalize villages by supporting cultural poverty alleviation. They participate in reconstruction of disaster stricken areas and ignite the light of hope for people in afflicted areas. They attend to art education and let more people benefit from arts. They are open-minded with a global view; they use their expertise to give back to society; they are willing to reach out to the world and take greater responsibility.

Back to life

Design originates from life and returns to life. Chinese architects know that their design will have a soul only if they make design relevant to life. Design expands possibilities of life and in return life bestows inspirations to design. Chinese architects are enthusiastic about life. They are open-minded to embrace the world. They care about arts and fashion, and broaden their fields to industrial design and costume design that requires innovation and creativity as well. They are gastronomes and carry forward Chinese dietetic culture that highlights health and science; they personalize their private spaces to show their life attitude; they have online interactive platform where they share and cooperate in design; they travel around to explore the world to get inspired, and they know how to enjoy life in their spare time.

中国建筑正在改变，改变的背后有中国社会快速转型和发展的必然性，而更主要的动因则源于中国建筑师的努力和推动。国家的高速发展给他们带来了史无前例的工程建设量，也为他们提供了施展才华的机会和实现梦想的平台。中国建筑师肩负着艰巨的使命，致力于城乡建设，促进着建筑和人居环境的和谐发展。他们不畏高强度的工作，坚持梦想，不忘初心；他们不断回溯和反思中国的建筑历程，并寻求改变和突破。

如今的中国建筑师在各个方面都有着我们想象不到的延展和尝试，他们回归建筑的本源，上下求索，实践创新；他们回归社会，用自己的力量推动着社会的发展；他们回归生活，让生活反哺设计，用设计点亮生活。

回归本源
面对快速发展带来的城市、建筑同质化等问题，中国建筑师积极面对，努力探寻解决之道，他们反思是否走得太快，是否丢失了什么，他们思考建筑的本质为何物，中国的城市和建筑将走向何方。在多样性的思考和可能性的实践中，中国建筑师们学习创新、内外融合，找到了开启中国建筑新篇章的钥匙，他们相信中国建筑转型的核心不是追随别人的脚步，而是倾听自己的声音，是回归建筑的本源，即根植于自然和地域、立足于历史和文化、专注于艺术和技艺。

回归社会
社会由无数差异化的个体汇聚而成，中国建筑师不仅在为普通大众创造更具价值的生活环境，更心系被忽略的人群和领域，从而承担起对于整个社会的责任。他们坚守在大都会，更新落后的边缘区域，改善弱势群体的生存环境，用设计的力量将爱传递至每一个普通的角落；他们扎根于村镇乡里，挽救即将失落的地域文化，力图实现乡村复兴；他们致力于公益事业，参与灾区的救助和重建，帮助灾区人民重拾生活的希望；他们关注艺术的普及和教育，让艺术的种子传播得更远更广。他们心怀天下，用看似微小的力量回报社会，他们努力走向世界，肩负起更大的责任。

回归生活
设计从生活中来，又回归到生活中去，中国建筑师深谙其中的道理，他们知道只有将设计与生活相融合，设计才会有灵魂，才能为生活创造更多可能性，生活也会回馈给设计无限的灵感。中国建筑师热爱生活，以开放和积极的姿态拥抱生活，他们关注艺术和时尚，用创意和革新向工业设计、服装设计等其他领域拓展；他们品味美食，承扬健康、科学的中华饮食文化；他们营造个性化的私人空间，分享自己的生活态度；他们打造线上设计交互平台，致力于网络端设计协作与分享；他们游历并探索未知世界，汲取不同地域的生命力与创造力，体会和创造工作之外的美好。

Appendix

附录

Arcplus Group PLC	Hongqiao International Airport T1 Renovation and GTC Project Design: 2012.02-2015.09 Construction: 2014.11-2016.10 Site Area: 103,227m² Floor Area: 203,746m²	Relics Park for the Coal Dock of Xiaguan Power Plant Design: 2012-2013 Construction: 2013-2014 Site Area: 9,200m² Floor Area: 3,922m²	The Restoration of Holy Trinity Church Design: 2005.05-2007.04 Construction: 2007.04-2010.04 Site Area: 3,500m² Floor Area: 2,265m²	D23 Project, Plot 8 of Hongqiao Business District Design: 2012.12 Construction: 2014.07 Site Area: 43,710m² Floor Area: 230,000m²
	Hexing Warehouse Renovation, 2010 Shanghai Expo Park, China Design: 2008.10-2009.6 Construction: 2009.6-2010.4 Site Area: 16,521m² Floor Area: 2,241m²	Preservation and Restoration of the Joint Trust Warehouse Design: 2014.07-2015.08 Construction: 2015 Floor Area: 25,550m²	Fairmount Peace Hotel Renovation and Expansion Project Design: 2007-2010 Construction: 2009-2010 Floor Area: 51,149m²	Customs Clearance Service Center of Chenglingji Free Trade Zone in Hunan Design: 2015 Construction: 2016 Site Area: 36,935m² Floor Area: 49,250m²
	WEI Retreat Tianmu Lake Design: 2013.9-2014.7 Construction: 2014.7-2016.4 Site Area: 7,385m² Floor Area: 8,998m²	Bund 33# renovation of the original British Consulate Building and the apartment Design: 2007-2010 Construction: 2009-2012 Site Area: 22,000m² Floor Area: 20,000m²	Preservation and Reparation Project of Shanghai Great World Design: 2004.10-2017.03 Construction: 2009.10-2017.03 Site Area: 31,893m² Floor Area: 16,626m²	Humble Administrator's Villa Design: 2010.05-2016.10 Construction: 2013-2016 Site Area: 42,326m² Floor Area: 35,072m²
	Yichang Planning Exhibition Hall Design: 2013.05-2014.12 Construction: 2016.02 Site Area: 29,983m² Floor Area: 20,960m²	Restoration of the Building of Shanghai Kunju Opera Troupe Design: 2009-2010 Construction: 2011-2013 Site Area: 2,335m² Floor Area: 3,293m²		
Beijing Institute of Architectural Design Co; Ltd.	Quadrangle Renovation, Caochang Area, Beijing(Yard 8,19,36,41) Design: 2015.03-2015.10 Construction: 2015.10-2016.08 Site Area: 4,360m² Floor Area: 3,160m²	Tianning NO.1 Culture and Creative Industrial Park Design: 2015-2016 Construction: 2016 Site Area: 7,900m² Floor Area: 80,000m²	Affiliated High School of Peking University Design: 2014-2015 Construction: 2014-2016 Floor Area: 26,000m²	Shenzhen Maritime Sports Base and Marine Navigation Sports School Design: 2009.04-2010.04 Construction: 2011.04 Site Area: 81,800m² Floor Area: 27,180m²
	Aimer Fashion Factory Design: 2004 Construction: 2013 Site Area: 62,236m² Floor Area: 53,000m²	Landscape and Related Facilities of The 9th China (Beijing) International Garden Expo Park Construction: 2011.01-2013.04 Site Area: 2,670,000m²		
China Architecture Design Group	Yushu Khamba Arts Center Design: 2011 Construction: 2014 Site Area: 24,563m² Floor Area: 20,610m²	Jixi Museum Design: 2009.11-2010.12 Construction: 2010.12-2013.11 Site Area: 9,500m² Floor Area: 10,003m²	Gymnasium of New Campus of Tianjin University Design: 2011.02-2013.08 Construction: 2015.11 Site Area: 33,950m² Floor Area: 18,798m²	Mogao Grottoes Digital Exhibition Center Design: 2008 Construction: 2014 Site Area: 40,000m² Floor Area: 10,440m²
	Zhudian Hoffmam Kiln Culture Center Design: 2014-2015 Construction: 2015-2016 Site Area: 9,777m² Floor Area: 1,650m²	Renovation of Countryside Buildings of Xibang Village Design: 2014-2015 Construction: 2015-2016 Site Area: 2,775m² Floor Area: 1,643m²		
China IPPR International Engineering Co., Ltd.	Preservation and Reparation Project of The Capital Cinema Design: 2014.07-2015.10 Construction: 2016.02 Site Area: 1,304m² Floor Area: 2,232m²	Protection and Renewal Plan of Shichahai Neighborhood in Beijing (2013-2030) Design: 2015.10 Construction: Being constructed Site Area: 589ha	Comprehensive Building and Academic Center of FAST Engineering Observation Base Design: 2013.06~2014.05 Construction: 2014.02~2016.10 Site Area: 5,860m² Floor Area: 2,680m²	
Architects & Engineers Co., Ltd of Southeast University	Library on the Quay Design: 2015.02-2015.05 Construction: 2015 Site Area: 853m² Floor Area: 517m²	Jizhaoying Mosque Design: 2009.02-2010.01 Construction: 2011.12-2014.01 Site Area: 661m² Floor Area: 1,307m²		

单位	项目1	项目2	项目3	项目4
华东建筑集团股份有限公司	虹桥国际机场T1航站楼改造及交通中心工程 设计时间：2012.02-2015.09 建造时间：2014.11-2016.10 用地面积：103227m² 建筑面积：203746m²	南京下关电厂运煤码头遗址公园 设计时间：2012-2013 建造时间：2013-2014 用地面积：9200m² 建筑面积：3922m²	基督教圣三一堂修缮工程 设计时间：2005.05-2007.04 建造时间：2007.04-2010.04 建筑面积：3500m² 建筑面积：2265m²	虹桥商务区8号地块D23项目 设计时间：2012.12 建造时间：2014.07 用地面积：43710m² 建筑面积：230000m²
	中国2010年上海世博会和兴仓库改造 设计时间：2008.10-2009.10 建造时间：2009.6-2010.4 用地面积：16521m² 建筑面积：2241m²	四行仓库修缮工程 设计时间：2014.07-2015.08 建造时间：2015 建筑面积：25550m²	和平饭店修缮与整治工程 设计时间：2007-2010 建造时间：2009-2010 建筑面积：51149m²	湖南城陵矶综合保税区通关服务中心 设计时间：2015 建造时间：2016 用地面积：36935m² 建筑面积：49250m²
	天目湖微酒店 设计时间：2013.9-2014.7 建造时间：2014.7-2016.4 用地面积：7385m² 建筑面积：8998m²	上海外滩源33#原英国领事馆及官邸历史建筑保护及再利用工程 设计时间：2007-2010 建造时间：2009-2012 用地面积：22000m² 建筑面积：20000m²	上海大世界修缮工程 设计时间：2004.10-2017.03 建造时间：2009.10-2017.03 用地面积：31893m² 建筑面积：16626m²	拙政别墅 设计时间：2010.05-2016.10 建造时间：2013-2016 用地面积：42326m² 建筑面积：35072m²
	宜昌规划展览馆 设计时间：2013.05-2014.12 建造时间：2016.02 用地面积：29983m² 建筑面积：20960m²	上海昆剧团大楼修缮改造工程 设计时间：2009-2010 建造时间：2011-2013 用地面积：2335m² 建筑面积：3293m²		
北京市建筑设计研究院有限公司	北京前门草厂片区四合院改造 设计时间：2015.03-2015.10 建造时间：2015.10-2016.08 用地面积：4360m² 建筑面积：3160m²	天宁一号文化科技创新园 设计时间：2015-2016 建造时间：2016 用地面积：7900m² 建筑面积：80000m²	北京大学附属中学 设计时间：2014-2015 建造时间：2014-2016 建筑面积：26000m²	深圳海上运动基地暨航海运动学校 设计时间：2009.04-2010.04 建造时间：2011.04 用地面积：81800m² 建筑面积：27180m²
	爱慕时尚工厂 设计时间：2004 建造时间：2013 用地面积：62236m² 建筑面积：53000m²	第九届中国（北京）国际园林博览会园区绿化景观及相关设施建设项目 建造时间：2011.01-2013.04 用地面积：2.67km²		
中国建筑设计院有限公司	玉树康巴艺术中心 设计时间：2011 建造时间：2014 用地面积：24563m² 建筑面积：20610m²	绩溪博物馆 设计时间：2009.11-2010.12 建造时间：2010.12-2013.11 用地面积：9500m² 建筑面积：10003m²	天津大学新校区综合体育馆 设计时间：2011.02-2013.08 建造时间：2015.11 用地面积：33950m² 建筑面积：18798m²	敦煌莫高窟数字展示中心 设计时间：2008 建造时间：2014 用地面积：40000 m² 建筑面积：10440m²
	祝甸砖窑文化馆 设计时间：2014-2015 建造时间：2015-2016 用地面积：9777m² 建筑面积：1650m²	西浜村农房改造工程 设计时间：2014-2015 建造时间：2015-2016 用地面积：2775m² 建筑面积：1643m²		
中国中元国际工程有限公司	首都电影院 设计时间：2014.07-2015.10 建造时间：2016.02 用地面积：1304m² 建筑面积：2232m²	什刹海街区保护与更新发展规划（2013-2030） 设计时间：2015.10 建造时间：建造中 用地面积：589hm²	FAST工程观测基地综合楼和学术交流支撑平台及附属设施工程 设计时间：2013.06-2014.05 建造时间：2014.02-2016.10 用地面积：5860m² 建筑面积：2680m²	
东南大学建筑设计研究院有限公司	码头书屋 设计时间：2015.02-2015.05 建造时间：2015 用地面积：853m² 建筑面积：517m²	吉兆营清真寺 设计时间：2009.02-2010.01 建造时间：2011.12-2014.01 用地面积：661m² 建筑面积：1307m²		

Firm	Project 1	Project 2	Project 3	Project 4
School of architecture at Southeast University; Architects & Engineers Co., Ltd. of Southeast University	Site Museum of Jinling Grand Bao'en Temple Design: 2011-2013 Construction: 2013-2015 Site Area: 75,300m² Floor Area: 60,800m²	Nanjing Yu Garden - A Project to Improve the Surrounding Environment Design: 2008.04-2014.04 Construction: 2014.04 Site Area: 34,500m² Floor Area: 3,618m²	Networking Engineering Center, Nanjing Sample Sci-Tech Park Design: 2010.07-2013.11 Construction: 2012.04-2014.07 Site Area: 24,600m² Floor Area: 21,505m²	
Southeast University School of architecture	Incremental Update Design of Gunan Street Historical Culture Districts in Yixing Design: 2012-2017 Construction: 2017 Site Area: 38,423m² Floor Area: 2,172m²	The Yinlu Tea House of Lianhuadang, Yixing Design: 2015.05-2015.11 Construction: 2015.11-2016.05 Site Area: 392m² Floor Area: 392m²		
Architectural Design and Research Institute of Tsinghua University Co.,Ltd.	Huashan Forum and Ecological Plaza Design: 2008.08-2009.11 Construction: 2011.04 Site Area: 408,008m² Floor Area: 8,867m²	Happiness Garden Exhibition Hall, Beichuan Earthquake Memorial Park Design: 2009.06-2010.04 Construction: 2010.12 Site Area: 2,308m² Floor Area: 2,353m²	Protection Plan Fujian Tulou, of the World Cultural Heritage Design: 2010.09-2013.02 Site Area: 10,870,000m²	Preservation Project of Prince Gong Mansion Design: 2004.10-2008.08 Construction: 2008.12 Site Area: 32,000m² Floor Area: 12,600m²
Tongji Architectural Design(Group) Co.,Ltd.	Shanghai Chess Institute Design: 2012.08-2013.02 Construction: 2013-2016 Site Area: 6,002m² Floor Area: 12,424m²	Power Station of Art Design: 2011.4-2011.9 Construction: 2011.9-2012.9 Site Area: 19,103m² Floor Area: 41,000m²	China Welfare Institute Pujiang Kindergarten Design: 2011.12-2014.03 Construction: 2013.06-2015.05 Site Area: 5,092m² Floor Area: 15,329m²	Liu Haisu Art Museum Design: 2012 Construction: 2015 Site Area: 6,000m² Floor Area: 12,540m²
	Shan dong Art Gallery Design: 2011.9-2012.9 Construction: 2012.1-2013.8 Site Area: 20,700m² Floor Area: 52,138m²			
Architecture Design institute of Huazhong university	The Dinosaur egg remainder museum in Qinglong Mountain Design: 2011 Construction: 2012 Site Area: 5,000m² Floor Area: 1,000m²			
CCTN Architectural Design Co.,Ltd.	2022 The Winter Olympics Plaza Design: 2016.03 Construction: 2016.04 Site Area: 76,952m²	Shougang Museum Design: 2016.08 Construction: 2017.03 Site Area: 72,844m²		
China Northwest Architecture Design & Research Institute Co., Ltd.	Xi'an South-Gate Plaza Improvement Project Design: 2011.01-2013.09 Construction: 2013.09-2014.09 Site Area: 82,879m² Floor Area: 69,229m²			
Zhejiang Greenton Architectural Design Co.,Ltd.	Historic Block Space Regeneration -Seclusive Jiangnan Boutique Hotel Design: 2016.01-2016.04 Construction: 2016 Site Area: 1,237m² Floor Area: 2,816m²	Contemporary Collective Living: New Forms of Affordable Housing for Relocalized Farmers in Hangzhou, China Design: 2014 Construction: 2016 Site Area: 19,277m² Floor Area: 15,286m²		
Tenio (Tianjin) Architecture and Engineering Co.,Ltd.	TENIO Green Design Center Design: 2011.11-2012.04 Construction: 2012.04-2012.12 Site Area: 3,215m² Floor Area: 5,756m²			

设计单位	项目1	项目2	项目3	项目4
东南大学建筑学院、东南大学建筑设计研究院有限公司	金陵大报恩寺遗址博物馆 设计时间：2011-2013 建造时间：2013-2015 用地面积：75300m² 建筑面积：60800m²	南京愚园地块保护与整治改善 设计时间：2008.04-2014.04 建造时间：2014.04 用地面积：34500m² 建筑面积：3618m²	南京三宝科技集团物联网工程中心 设计时间：2010.07-2013.11 建造时间：2012.04-2014.07 用地面积：24600m² 建筑面积：21505m²	
东南大学建筑学院	宜兴市古南街历史文化街区渐进式更新设计 设计时间：2012-2017 建造时间：2017 用地面积：38423m² 建筑面积：2172m²	隐庐莲舍：宜兴莲花荡农场茶室 设计时间：2015.05-2015.11 建造时间：2015.11-2016.05 用地面积：392m² 建筑面积：392m²		
清华大学建筑设计研究院有限公司	华山论坛及生态广场 设计时间：2008.08-2009.11 建造时间：2011.04 用地面积：408008m² 建筑面积：8867m²	北川抗震纪念园幸福园展览馆工程 设计时间：2009.06-2010.04 建造时间：2010.12 用地面积：2308m² 建筑面积：2353m²	世界文化遗产福建土楼保护规划总纲 设计时间：2010.09-2013.02 用地面积：10.87km²	恭王府府邸文物保护修缮工程 设计时间：2004.10-2008.08 建造时间：2008.12 用地面积：32000m² 建筑面积：12600m²
同济大学建筑设计研究院（集团）有限公司	上海棋院 设计时间：2012.08-2013.02 建造时间：2013-2016 用地面积：6002m² 建筑面积：12424m²	上海当代艺术博物馆 设计时间：2011.4-2011.9 建造时间：2011.9-2012.9 用地面积：19103m² 建筑面积：41000m²	中福会浦江幼儿园 设计时间：2011.12-2014.03 建造时间：2013.06-2015.05 用地面积：5092m² 建筑面积：15329m²	刘海粟美术馆 设计时间：2012 建造时间：2015 用地面积：6000m² 建筑面积：12540m²
	山东美术馆 设计时间：2011.9-2012.9 建造时间：2012.1-2013.8 用地面积：20700m² 建筑面积：52138m²			
华中科技大学建筑设计研究院	青龙山恐龙蛋遗址博物馆 设计时间：2011 建造时间：2012 用地面积：5000m² 建筑面积：1000m²			
中联筑境建筑设计有限公司	2022首钢西十冬奥广场 设计时间：2016.03 建造时间：2016.04 用地面积：76952m²	首钢博物馆 设计时间：2016.08 建造时间：2017.03 用地面积：728442m²		
中国建筑西北设计研究院有限公司	西安南门广场综合提升改造项目 设计时间：2011.01-2013.09 建造时间：2013.09-2014.09 用地面积：82879m² 建筑面积：69229m²			
浙江绿城建筑设计有限公司	历史街区空间再生 ——隐居江南精品酒店 设计时间：2016.01-2016.04 建造时间：2016 用地面积：1237m² 建筑面积：2816m²	乡村低收入住宅 ——杭州富阳东梓关回迁安置农居 设计时间：2014 建造时间：2016 用地面积：19277m² 建筑面积：15286m²		
天友（天津）建筑设计股份有限公司	天友绿色设计中心 设计时间：2011.11-2012.04 建造时间：2012.04-2012.12 用地面积：3215m² 建筑面积：5756m²			

Firm	Project 1	Project 2	Project 3
CITIC General Institute of Architectural Design and Research Co., Ltd.	Creative Valley of South Taizi Lake (Phase I) Design: 2012-2013 Construction: 2014-2017 Site Area: 31,816m² Floor Area: 56,900m²	Comprehensive Renovation Project of Zhongshan Road (from Jianghan Road to Yiyuan Road) Design: 2015.04 Construction: 2016.05 Site Area: 213,350m²	Planning of Xiedian Traditional Village Protection and Regeneration Design: 2015.07-2016.01 Construction: 2016.01-2016.09 Site Area: 110 mu Floor Area: 9,400m²
Tianjin Architecture Design Institute	Shooting Range Hall of the East Asian Games Design: 2010-2011 Construction: 2011-2013 Site Area: 74,811m² Floor Area: 37,996m²	Tianjin TV Station Design: 2006-2012 Construction: 2008-2016 Site Area: 246,000m² Floor Area: 260,000m²	
Central-South Architectural Design Institute Co., Ltd.	Xi'an North Station of Zhengzhou-Xi'an High-speed Railway Design: 2008.09-2009 .07 Construction: 2008-2010 Site Area: 439,043m² Floor Area: 332,000m²		
On-earth Architecture	Macha Village Community Center Design: 2013.05-2014.04 Construction: 2014.06-2016.05 Site Area: 1,860m² Floor Area: 648m²	Mendeng Village Community Center Design: 2015.06-2015.10 Construction: 2015.08-2015.10 Site Area: 257m² Floor Area: 213m²	
Beijing Tsinghua TongHeng Urban palnning& Design Institute	THE-Studio (Tsinghua Eco Studio) Design: 2015.02-2015.04 Construction: 2015.06 Site Area: 1,826m² Floor Area: 701m²		
China Southwest Architecture Design & Research Institute Co.,Ltd.	Homeland of Mosuo, The Project of Protecting Mosuo Habitation Design: 2014 Construction: 2015		
Amateur Architecture Studio	Renovation of Wencun Village Design: 2013.01-2015.06 Construction: 2014.07-2015.10 Site Area: 48,905m² Floor Area: 7,590m²		
Faculty of Architecture and City Planing, KUST; Architecture Institute of Vulgar Autochthonous, KUST	Hani Nationality Mushroom House Retrofit Experiment in AZheKe Village, Yuanyang County Design: 2015.05 Construction: 2015.07-2015.11 Site Area: 120m² Floor Area: 245m²		
Institute of Architecture Design & Planning Co., Ltd, Nanjing University	A Rural Shop at Huashu Village, Nanjing Design: 2014.09 Construction: 2015.06 Site Area: 668m² Floor Area: 150m²		
United Design Group Co. Ltd., Shanghai	Oriental Rural Life Pavilion, Huishan, Wuxi Design: 2013 Construction: 2013 Site Area: 1,500m² Floor Area: 1,337m²		
Atelier Heimat	Lounge Bridge Renovation Design: 2016.05-2016.07 Construction: 2016.09-2017.01 Site Area: 320m² Floor Area: 193m²		
Swiss Federal Institute of Technology (ETHZ)	Shaxi Rehabilitation Project Design: 2004 Construction: 2004-2010 Site Area: 8,150m² Floor Area: 3,200m²		
Zhejiang Ancient Architecture Design and Research Institute	The Practice of Protection and Utilization of Traditional Villages in Yangjiatang Village , Songyang County Design: 2013 Construction: 2014		
WangLei Architects	Wooden Bridge Design: 2015.04 Construction: 2015.09 Site Area: 200m² Floor Area: 150m²		
Evolution Design	Renovation Project of Grandma's Yard in Huangshandian Village Design: 2016.4 Construction: 2016.9 Site Area: 720m² Floor Area: 305m²		

设计单位	项目1	项目2	项目3	项目4
中信建筑设计研究总院有限公司	武汉南太子湖创新谷（一期） 设计时间：2012-2013 建造时间：2014-2017 用地面积：31816m² 建筑面积：56900m²	中山大道综合改造工程（江汉路至一元路） 设计时间：2015.04 建造时间：2016.05 用地面积：213350m²	谢店村传统村落保护与再生规划设计 设计时间：2015.07-2016.01 建造时间：2016.01-2016.09 用地面积：110亩 建筑面积：9400m²	
天津市建筑设计院	东亚运动会射击馆 设计时间：2010-2011 建造时间：2011-2013 用地面积：74811m² 建筑面积：37996m²	天津电视台 设计时间：2006-2012 建造时间：2008-2016 用地面积：246000m² 建筑面积：260000m²		
中南建筑设计院股份有限公司	郑西高铁西安北站 设计时间：2008.09-2009.07 建造时间：2008-2010 用地面积：439043m² 建筑面积：332000m²			
土上建筑工作室	马岔村村民活动中心 设计时间：2013.05-2014.04 建造时间：2014.06-2016.05 用地面积：1860m² 建筑面积：648m²	门等村村民活动中心 设计时间：2015.06-2015.10 建造时间：2015.08-2015.10 用地面积：257m² 建筑面积：213m²		
北京清华同衡规划设计研究院有限公司	清控人居科技示范楼 设计时间：2015.02-2015.04 建造时间：2015.06 用地面积：1826m² 建筑面积：701m²		中国建筑西南设计研究院有限公司	摩梭家园——摩梭人聚居地保护 设计时间：2014 建造时间：2015
业余建筑工作室	文村村新建民居 设计时间：2013.01-2015.06 建造时间：2014.07-2015.10 用地面积：48905m² 建筑面积：7590m²		昆明理工大学建筑与城市规划学院、昆明本土建筑设计研究所有限公司	元阳县阿者科村哈尼族蘑菇房改造实验 设计时间：2015.05 建造时间：2015.07-2015.11 用地面积：120m² 建筑面积：245m²
南京大学建筑规划设计研究院有限公司	南京桦墅村村口乡村铺子 设计时间：2014.09 建造时间：2015.06 用地面积：668m² 建筑面积：150m²		上海联创建筑设计有限公司	无锡惠山东方田园生活馆 设计时间：2013 建造时间：2013 用地面积：1500m² 建筑面积：1337m²
合木建筑工作室	驿道廊桥改造 设计时间：2016.05-2016.07 建造时间：2016.09-2017.01 用地面积：320m² 建筑面积：193m²		瑞士联邦理工大学	沙溪复兴工程 设计时间：2004 建造时间：2004-2010 用地面积：8150m² 建筑面积：3200m²
浙江省古建筑设计研究院	松阳县杨家堂村传统村落保护与发展实践 设计时间：2013 建造时间：2014		王磊联合建筑师事务所	木桥听雨 设计时间：2015.04 建造时间：2015.09 用地面积：200m² 建筑面积：150m²
空间进化（北京）建筑设计研究院有限公司	黄山店姥姥家民宿改造项目 设计时间：2016.04 建造时间：2016.09 用地面积：720m² 建筑面积：305m²			

Shen Di

1982, Graduate from Tongji University, Master's Degree
National Engineering Survey and Design Master
Professor-level Senior Engineer
First-level Registered Architect
Vice-president and Chief Architect of Arcplus Group PLC
Vice-president, Architects Branch of the Architectural Society of China

Representative Works and Awards
Planning for World Expo 2010 Shanghai
Shanghai World Financial Center Project
COSCO Two Bay City Planning and Project
Expo Axis and Underground Complex
Shanghai Dongjiao State Guest Hotel
Renovation and Expansion Project of Jingxi Hotel
Expo Bao Steel Grand Stage
Mine Ecological Rehabilitation and Utilization Project of Dawang Mountain in Changsha
Renovation Project of Shanghai Shendu Mansion
Renovation Project of Shanghai Eco-Home

沈迪

1982 年毕业于同济大学建筑系，硕士学位
全国工程勘察设计大师
教授级高级工程师
国家一级注册建筑师
华东建筑集团股份有限公司副总裁，总建筑师
中国建筑学会建筑师分会副理事长

主要代表作品：
2010 年上海世博会规划
上海环球金融中心
中远两湾城规划及一期项目
世博轴及地下综合体工程
上海东郊宾馆
京西宾馆整体改扩建项目
世博宝钢大舞台
长沙大王山矿坑生态修复利用工程
上海申都大厦改造工程
沪上生态家改造工程

The Architectural Society of China

The Architectural Society of China (ASC) is a national academic institute constituted by the professionals of architectural science and technology in China. It is the member institute of China Association of Science and Technology and closely related by profession to the Ministry of Housing and Urban-Rural Development.

ASC was founded in October 1953. Mr. Liang Sicheng, Mr. Yang Tingbao and other famous Chinese architects assumed the leadership in the following years. ASC convenes the most outstanding experts, scholars and engineering technicians of the architectural field in China, and becomes the reliable think-tank and assistant of the government in promoting the urban and rural development.

The membership of the ASC consists of individual members (honorary members, fellows, members, student members and correspondence members) as well as corporation (units) members. ASC has over 10,000 individual members and over 300 corporation members. ASC entrusts its sub-institutes and local chapters to assist for admitting members and practises dependency administration.

The periodicals published by ASC and its sub-institutes are: Architectural Journal, Journal of Building Structures, Architectural Knowledge (a+a), Building Economy, Heating Ventilation and Air-conditioning, Building Heating and VA, Engineering Surveying, Aseismatic Engineering, Asian Architectural Water Supply and Drainage Institute, etc. The official ASC website is www.chinaasc.org.

ASC joined in the International Union of Architects as a national section in 1955 and joined in the Architects Regional Council of Asia as a national member in 1989. ASC also established and maintained academic exchange and friendly cooperative relationship with Royal Institute of British Architects, American Institute of Architects, Architectural Institute of Japan, Union of Architects of Russia, Architectural Institute of Korea, Hong Kong Institute of Architects, Union of Architects of Mongolia, Hungarian Institute of Architects, Polish Institute of Architects, etc.

In order to support and encourage the work and contribution of the architectural engineering professionals, ASC awards over dozen prizes including China Architectural Design Award and Liang Sicheng Architecture Medal which is the supreme honor in the Chinese architectural field.

中国建筑学会

中国建筑学会是全国建筑科学技术工作者组成的学术性团体，主管单位为中国科学技术协会、住房和城乡建设部。

学会于1953年10月23日成立。著名建筑专家梁思成、杨廷宝先生历任本会领导人。学会集中了中国建筑界各专业最优秀的专家、学者和工程技术人员，成为政府推进城乡建设最可靠的智囊和助手。

学会设有个人会员（包括会员、资深会员、名誉会员、学生会员、外籍会员）和团体会员。目前本会的个人会员有10万余人，团体会员300多家。本会委托地方学会协助发展会员，并实行属地管理。

学会编辑出版的刊物有《建筑学报》、《建筑结构学报》、《建筑知识》，以及《建筑经济》、《暖通空调》、《建筑热能与通风空调》、《工程抗震》、《工程勘察》、《亚洲建筑给水排水》等。中国建筑学会官网为：www.chinaasc.org

中国建筑学会在1955年以国家会员身份加入了国际建筑师协会（UIA），1989年以国家会员身份加入了亚洲建筑师协会（ARCASIA）。此外，与英国皇家建筑师学会、美国建筑师学会、日本建筑学会、俄罗斯建筑师联盟、韩国建筑学会、中国香港建筑师学会、匈牙利建筑学会、波兰建筑学会等数十个国家相关组织建立学术交流和友好合作关系。

学会设有"梁思成建筑奖"、"中国建筑设计奖"等十余奖项，以支持和鼓励中国建筑界工程技术人员的工作和贡献。

Architects Branch of Architectural Society of China

The Architects Branch is the highest academic group with well-known experts and scholars in the field of architectural design, since its inception in 1989, in accordance with the purpose of "branch to carry out academic activities, improve the architect's theory and practice level, flourish architectural creation; play a role in the society of architects, the architect of the maintenance of rights and interests; and pay attention to the cultivation of talents to improve the construction architectural education, communication and cooperation and the development of architects all over the world". A lot of work has been completed for more than 20 years, and it has played an important role in promoting the progress of the industry.

According to the different academic fields, the architects branch has 16 specialized committees, respectively, architectural creation and education special committee, hospital building committee, education construction committee, building technology committee, environmental art committee, human settlements special committee, architectural arts committee, architectural photography association, mural art committee, western building committee, green building committee, digital architectural design committee, international high-rise building exchange committee, construction planning committee, regional construction committee, and rural construction committee.

中国建筑学会建筑师分会

中国建筑学会建筑师分会是中国建筑师的最高学术团体，集中了全国建筑设计领域的知名专家、学者，自1989年成立至今，按照分会"开展学术活动，提高建筑师的理论水平和实践水平，繁荣建筑创作；发挥建筑师的社会作用，维护建筑师的权益；关注建筑教育建筑人才的培养和提高；发展与世界各国建筑师的交流与合作"的宗旨，在20多年里做了大量的工作，起到了推动行业进步的重要作用。

根据学术领域的不同，建筑师分会下设16个专业委员会，分别是建筑创作与理论专委会，医院建筑专委会，教育建筑专委会，建筑技术专委会，环境艺术专委会，人居环境专委会，建筑美术专委会，建筑摄影专委会，壁画艺术专委会，西部建筑专委会，绿色建筑专委会、数字建筑设计专委会、高层建筑国际交流专委会、建筑策划专委会、地区建筑专委会以及乡村建筑专委会。

Arcplus Group PLC

Arcplus Group PLC (hereinafter referred to as "Arcplus") is an architecture-based and advanced technology service provider. With engineering design and consulting as its core business, Arcplus holds its position as an integrated service supplier offering high quality comprehensive solutions for urban construction. It has more than 10 subsidiary design firms and studios, providing a full spectrum of services, e.g. pre-stage consultation, architectural and engineering design, water engineering, landscape and interior design. For more than 10 consecutive years, it has been ranked by the US magazine "Engineering News Record" (ENR) as one of "The Top 150 Global Design Firms."

Talents

Arcplus recognizes and treats our employees as the most valuable and treasured assets. At present, Arcplus has more than 7000 talented individuals in disciplines of various specialization, comprising of 2 academicians of China Academy of Engineering, 6 China National Survey and Design Masters, 1320 senior professionals, and 1600 certified engineers in architecture, planning, engineering and quantity surveying. Our dedicated team specializes in conceptualizing, planning, designing and managing a myriad of projects, from schools to high-rise commercial complexes, hospitals, residential developments, airports, and transportation infrastructures.

Services

Arcplus prides themselves on their abilities to provide a one-stop service solution and a complete range of multi-disciplinary services that their clients need. Their services encompass the whole lifecycle of a project, including urban planning and design, architectural design, water engineering, municipal engineering, landscape design, interior design, site survey, building acoustic design, and EPC. Arcplus has been granted by China Association for Quality the award of "Customer Satisfied Enterprise" and "Customer Satisfied Service Provider".

Achievements

With a strong track record of more than 10 thousand design and consulting projects, Arcplus today has expanded its portfolio to 29 provinces across China and more than 20 countries and regions worldwide. Their quality architectural and engineering design has generated some of China's most significant and enduring landmarks. Arcplus has established 1 national technology center and 5 Shanghai municipal engineering technology research centers. More than 1900 design projects, scientific research programs, and standards and codes compiled have been granted national, provincial or municipal awards for outstanding design and technological advancement and innovation. They have led or participated in compilation for more than 240 standards and codes of various disciplines on national, industry or municipal level, and obtained 410 intellectual properties.

华东建筑集团股份有限公司

华东建筑集团股份有限公司是一家以建筑设计为核心、以先瞻科技为依托的技术服务型上市企业，集团定位为以工程设计咨询为核心，为城镇建设提供高品质综合解决方案的集成服务供应商。旗下拥有华东建筑设计研究总院、上海建筑设计研究院、华东都市建筑设计研究总院、工程建设咨询公司、上海水利工程设计研究院、建筑装饰环境设计研究院、美国威尔逊室内设计公司等十余家分子公司和专业机构。连续十多年被美国《工程新闻纪录》（ENR）列入"全球工程设计公司150强"企业。

历史与品牌　1952年5月19日，华东工业部建筑设计公司成立；次年1月，上海市建筑工程局生产技术处成立设计科；1998年，"两院"合并组建集团，成为国内建筑设计行业第一家完全走向市场的大型企业集团公司。2015年10月30日，集团由国有独资公司转变为国有控股的上市公司。

团队与人才　集团技术力量雄厚，人才梯队合理，现有各类专业员工7000余人，其中中国工程院院士2名、全国勘察设计大师6名、上海市企业领军人才7名，教授级高级职称130余人、高级职称1200余人，拥有注册建筑师、注册规划师、注册工程师、注册造价师等各类专业注册从业资格的人员1600余人。

服务与质量　集团业务领域覆盖工程建设项目全过程，其中包括规划、建筑、水利、市政、风景园林、室内装饰、岩土、建筑声学等各类设计咨询服务，以及设计、采购、施工一体化（EPC）工程总承包服务。集团连续多年荣获中国质量协会颁发的"用户满意企业"奖和"全国用户满意服务"奖。

经验与成就　集团作品遍及全国29个省、自治区、直辖市及20多个国家和地区，累计完成上万项工程设计及咨询工作，建成大量地标性项目。集团拥有1个国家级企业技术中心和5个上海市工程技术研究中心，超过1900项工程设计、科研项目和标准设计荣获国家、省（市）级优秀设计和科研进步奖，主持和参与编制了各类国家、行业及上海市规范、标准共240余册，获得知识产权410余项。

市场与合作　集团长期与各类国际顶级设计机构建立广泛联系，与政府、开发区、金融机构、地产、文化机构等各类社会资源密切合作。在北京、重庆、大连、武汉、深圳、广州等23个城市共设立有26个常驻分支机构，在香港设有全资子公司，全资控股子公司美国威尔逊室内设计公司在达拉斯、纽约、洛杉矶等全球8个城市设有办公室。

文化与理念　传承建筑文化，以完美的创意和先进的建筑技术，将古典与现代、艺术与商业、舒适与功能、美感与科技完美地结合在一起，秉持绿色设计价值观，赋予建筑独特的风格和超凡的魅力，最大程度地满足客户的需求，推动未来城市的可持续发展。

梦想与追求　我们出身建筑专业，信仰至臻设计，用"设计"的思维，严谨地规划每一个细节，实现对极致的追求。秉承这一初心，在更广阔的领域里，我们衍生服务内容，从设计到咨询、信息、科研、投资……创意成就梦想，设计构筑未来。

Beijing Archicity Consulting Co., Ltd

Beijing Archicity Consulting Corp (ARCHICITY) is a business consulting firm helping clients understand, adapt to and prosper from the rapidly changing social, political , technological, and business drivers in the architecture and planning industry.

Having worked with the Chinese government departments in urban planning and architecture industry and advised some of the largest and most sophisticated corporate, industrial, and NGOs in the world for over 18 years, we combine insider policy analysis with deep construction industry knowledge to offer unparalleled insights and cutting edge strategies maximizing business opportunities.

For over 18 years, ARCHICITY have coordinated 70,000 Chinese architects and planners for global outreach in academic exchange and professional practice. Within all the practices, ARCHICITY has participated in more than 1000 consulting cases.

北京凯欣城市发展咨询有限公司

北京凯欣城市发展咨询有限公司是一家服务建筑、规划行业的咨询公司。 我们的专业团队帮助客户深入了解并且适应中国的建筑、规划市场。在中国快速发展的社会、经济、技术和商业环境下为客户提供最优解决方案。

在过去的 15 年里，凯欣参与了许多政府项目的咨询，我们的咨询顾问团队帮助国有企业、行业协会与学会、政府机构和私企开发集团提供市场与职业发展建议以及第三方技术支持。 我们的咨询顾问团队具有客观有效的政策分析方法和深厚的行业知识，为客户拓展新的市场与项目提供了有效解决方案。

从 1999 年到 2017 年凯欣组织了超过 70000 名建筑及规划师参与海外学术交流和职业实践，策划了（包括直接与间接参与的）近 1000 个咨询项目。